Springer Aerospace Technology

The series explores the technology and the science related to the aircraft and spacecraft including concept, design, assembly, control and maintenance. The topics cover aircraft, missiles, space vehicles, aircraft engines and propulsion units. The volumes of the series present the fundamentals, the applications and the advances in all the fields related to aerospace engineering, including:

- structural analysis,
- aerodynamics,
- aeroelasticity,
- aeroacoustics,
- flight mechanics and dynamics
- orbital maneuvers,
- avionics,
- systems design,
- materials technology,
- launch technology,
- payload and satellite technology,
- space industry, medicine and biology.

The series' scope includes monographs, professional books, advanced textbooks, as well as selected contributions from specialized conferences and workshops.

The volumes of the series are single-blind peer-reviewed.

To submit a proposal or request further information, please contact: Mr. Pierpaolo Riva at pierpaolo.riva@springer.com (Europe and Americas) Mr. Mengchu Huang at mengchu.huang@springer.com (China)

The series is indexed in Scopus and Compendex

Anzhela Borzova · Yuri Chinyuchin ·
Vadim Vorobyov · Dmitry Zatuchny

Modeling and Forecasting of Staffing in Civil Aviation

Springer

Anzhela Borzova
Moscow State Technical University of Civil Aviation (MSTU CA)
Moscow, Russia

Yuri Chinyuchin
Moscow State Technical University of Civil Aviation (MSTU CA)
Moscow, Russia

Vadim Vorobyov
Moscow State Technical University of Civil Aviation (MSTU CA)
Moscow, Russia

Dmitry Zatuchny
Moscow State Technical University of Civil Aviation (MSTU CA)
Moscow, Russia

ISSN 1869-1730 ISSN 1869-1749 (electronic)
Springer Aerospace Technology
ISBN 978-981-19-6240-0 ISBN 978-981-19-6238-7 (eBook)
https://doi.org/10.1007/978-981-19-6238-7

This Springer imprint is published by the registered company Springer Nature Singapore Pte Ltd.
The registered company address is: 152 Beach Road, #21-01/04 Gateway East, Singapore 189721, Singapore

Introduction

The vastness of Russia's territory, the geographical and climatic features of remote economic regions, and the uneven distribution of business and cultural centers create objective prerequisites for the development of air transportation.

One of the documents, regulating the development of air transport in the Russian Federation in the future, is the state program "Development of the transport system" for 2016–2021 years.

Among the primary tasks of the development of civil aviation are noted: improving the efficiency of meteorological, air navigation and ground support that meet the modern requirements of scientific and technological progress.

At the same time, there is a need to reorient the system of industry education, taking into account the requirements of implementing innovative solutions for the development of the air transport industry.

In the works of Akinshin R. N., Bolyasnikov V. V., Bachkalo B. I., Bolshedvorskaya L. G., Eliseev B. P., Elisov L. N., Kovalenko G. V., Smurov M. Yu., Filin A. D., Sharov V. D., Shmelkov A. V., the scientific foundations of improving the efficiency of air transport operation are presented, which address the problems of the quality of personnel support, depending on the level of personnel training in educational organizations of the industry and their management with a focus on solving promising problems. To improve the efficiency of management in organizational and educational systems, it is possible to use invariant methods, models and mechanisms, considered in the works of A. M. Bershadsky, V. N. Burkov, M. B. Guzairov, B. P. Eliseev, V. A. Zernov, B. G. Litvak, Ya. E. Lvovich, D. A. Novikov, Yu. S. Sakharov, N. A. Selezneva.

However, the research, conducted to improve the human potential of civil aviation, is mainly based on qualitative analysis and logical mechanisms for managing the training system. On the other hand, invariant approaches to the management of organizational systems require a problem-oriented approach to the processes of air transport operation and the works that take into account the specifics of the industry in the formalized descriptions of educational processes cover only the local components of an integrated personnel support system. The lack of a unified methodology in theoretical research leads to the unsolved problems of improving the efficiency of personnel management of air transport enterprises. A systematic assessment of the ways of modernization of the branch network of educational organizations is required, which allows taking into account the prospects for the development of air transport, using mathematical methods of modeling and optimization.

Thus, there is a need to develop a modern methodological base of scientific theoretical and applied research, aimed at improving the processes of training aviation personnel, through a systematic analysis and synthesis of promising areas and tasks for the development of civil aviation, the problems of modernization of the network of branch educational organizations, the content of training to achieve, based on the use of forecasting and optimization models, compliance with the requirements for personnel support and the requirements for improving the efficiency of air transport operations.

In Chapter 1 of the monograph, a systematic analysis of the problems of managing the training of personnel with higher and secondary professional education in the field of air transport operation is carried out. The author characterizes innovative approaches, focused on the implementation of the strategy of development of civil aviation. The connection between innovations in the operation of air transport and the advanced requirements of dual training in the process of training specialists for air transport enterprises are established. Against this background, the features of personnel training in the field of air transport operation are formulated, depending on the areas of training, the requirements of educational programs and professional standards for secondary vocational and higher education. A new system of continuous professionalization of personnel through the use of a network approach to the organization of interaction of educational institutions in the industry is formulated. The peculiarity of this approach is that the formation of interaction between educational institutions of various types and levels in the format of vertically integrated ones will increase the role of specialized training centers and complexes of the air transport industry in various regions of Russia.

The results of the monitoring assessment of the effectiveness of the activities of educational organizations of civil aviation are considered. The possibility of using statistical material both on air transport efficiency indicators and the system of personnel training in this area for forecasting tasks and together with expert evaluation for optimization tasks is shown. A comparative analysis of a number of key indicators of effectiveness of universities that train specialists for all types of transport systems is carried out, which characterizes certain advantages of the cluster of educational organizations of civil aviation.

A systematic analysis of the problems of managing for the training of personnel with professional and higher education in the field of air transport shows that the basic mechanisms for improving the level of the system of personnel support are:

consideration of innovative and optimization approaches, focused on advanced requirements for the efficiency, quality and reliability of the functioning and development of air transportation in educational assets;

ensuring compliance of the volume characteristics of the system for personnel training and the requirements of air companies;

achieving maximum convergence of the competence characteristics of professional and educational standards.

As an invariant of the implementation of these mechanisms, it is proposed to consider a number of models that, on the one hand, allow for a reliable forecast of the need for personnel and, on the other hand, determine the general basis for the characterization of labor functions, competencies and educational assets. This message is the main essence of the methodology of the study of the system of personnel support formation, using a model-oriented approach. Such an approach in this area with a focus on the management of personnel training has not been used and requires, first of all, a integral conceptual foundation in the form of a number of principles.

In Chapter 2 of the book, the conceptual foundations of a model-based approach to forecasting and optimizing of the system for personnel training in the field of air transport operation are proposed. It is shown that they are based on a number of principles.

Principle 1—duality and multi-method use of models. This principle defines two directions of the model-oriented approach.

The first of them is related to the use of mathematical models for solving forecasting and optimization problems. In this case, we will divide the models into three groups:

predictive one;
estimated one;
optimization one.

In relation to the problem of improving the system of personnel training, predictive models allow us to determine the values of the volume characteristics of the need for personnel of air companies and training of specialists in universities on the basis of statistical information. Evaluation models are based on expert information and set priorities for the components of the modernization of the personnel training system. The optimization models integrate formalized information that allows us to link the optimized variables with extreme and boundary requirements for the effectiveness indicators of the system for personnel training.

At the same time, optimization modeling is a partial implementation of the duality principle in several aspects. The first aspect is determined by two mechanisms for achieving the effectiveness of the system for personnel training: reduction and transformation.

Under the reduction mechanism, we will understand the process of dividing the studied sets that characterize the system of personnel training into two subsets, one of which is more consistent with the extreme and boundary requirements for effectiveness indicators. Another form of reduction is the division of these sets within the framework of dual study, which provides a balance of supply and demand in the civil aviation labor market. In some cases, it is necessary to perform several reduction cycles. This mechanism will be called multi-stage reduction.

The transformation mechanism is aimed at forming such a set of elements of the personnel training system that the best show the properties of other sets, defined by regulatory documents.

The general structure of optimization models that show the mechanisms under consideration is related to another aspect of the implementation of the duality principle by dividing the indicators of effectiveness of the system for personnel training into two groups that meet extreme and boundary requirements.

The next aspect is related to two mechanisms for organizing the optimization modeling process: randomization and smoothing.

The second direction is focused on creating a bank of professional activity models, on the one hand related to the characterization of labor functions in professional standards, and on the other—showing in a formalized form the components of educational assets that are aimed at selecting of certain competencies of personnel in the field of air transport operation.

The models of the first and second directions can be implemented by different methods, which complement the principle of duality of the model-oriented approach with the characteristic of multimethodality. The multi-method approach to the application of models makes it necessary to conduct a comparative analysis of the results of solving problem-oriented problems.

Principle 2—expert-statistical formation of predictive and evaluation models. When justifying the first principle, it is indicated that statistical (retrospective) information is used to build predictive models. The main form of statistical information that determines changes in the system of personnel training is the time series of volume indicators of the labor market demand for specialists with higher education and graduation from universities.

To build a predictive model, it is proposed to use a number of time series processing methods that are typical for technical systems. The multi-method approach makes it possible to calculate predictive estimates with different quality indicators.

After obtaining the results of forecasting, using each of these methods, you should choose the result that most reliably shows the trends of change in the volume indicators in the system of personnel training for air companies, operated of air transport.

In this case, an evaluation model is built on the basis of expert information and the method that has the highest priority for this model is selected.

Principle 3—multi-alternative choice of professional activity models.

Professional activity is considered through the following categories: types of activity, labor functions, professional competencies and corresponding models of professional activity. The content asset of professional activity models is proposed to be formed in the form of structural, situational, functional and optimization descriptions.

Principle 4—optimal transformation of professional activity models into components of educational programs.

This principle is based on an optimization model that links multiple models of professional activity with multiple components of educational programs.

The task of transformation is to effectively show the elements of a variety of models in the formation of elements of a variety of educational programs.

In the case of dual training, which requires the separation of theoretical and practice-oriented components, multi-stage reduction and transformation are carried out using a tree-like scheme.

Chapter 3 of this book is devoted to the formation of predictive and evaluation models of the planned needs and activities of air company personnel. It is shown that it is possible to generate spatial and temporal information, based on statistical data, linked on the one hand to calendar periods, and on the other—to the geo-location of airports in the Russian Federation, using geographic information system (GIS). At the same time, one part of the information support characterizes the efficiency of air transport by a number of indicators: cargo turnover, passenger turnover, passenger traffic volume, the number of passengers, served by airports, the degree of renewal of the aircraft fleet; the second—the system of personnel training: indicators for monitoring the effectiveness of industry educational organizations and other information systems of the Ministry of Education and Science of Russia. The particular interest from the second part is the indicators of the need for specialists and graduation in the areas and specialties in the field of air transport maintenance and the cartographic asset for the employment of graduates. Using a multi-method approach, predictive models are constructed for the listed indicators.

It is established that in cases of GIS-oriented information it is possible to estimate the volume of employment of graduates of industry-specific educational organizations, depending on the location of airports. This analysis allows us to identify those air companies, for which it is necessary to optimize the mechanisms for managing their staffing support.

It is proved that the results of predictive modeling provide an opportunity for expert evaluation of the prospective values of the volume indicators of the system for personnel training in the field of air transport maintenance.

The process of forecasting is completed by geographically distributed planning of employment of graduates of aviation universities on the basis of GIS-oriented information.

Long-term planning of the system for personnel training is associated not only with the choice of volume indicators, but also with the definition of priority activities of air transport maintenance specialists, focused on the industry development strategy. To solve the second problem, which is the first step in the implementation of multi-stage reduction, it is proposed to implement evaluation modeling, based on expert information.

Chapter 4 presents the results of optimizing personnel training, based on the use of invariant models of professional activity in the field of air transport enterprise management. In the case of a model-based approach, it is necessary to organize a multi-stage reduction of the components of professional activity in the following sequence:

(1) reduction of the original set of directions;
(2) reduction of models of professional activity, corresponding to the directions, chosen at the first stage;
(3) reduction of invariant descriptions of professional activity models, selected at the second stage.

The first stage of optimization consists in expert selection of promising areas of management.

Optimization modeling as an invariant content of the basic tasks of managing the process of air transport maintenance within the air company consists of three components:

meaningful description of the management task;
creating an optimization model;
construction of a procedure that allows you to find a solution of an optimization problem.

The method of balanced scorecard is used as the basis for the content formulation of the optimal choice problem, which provides the ability to coordinate different levels of company management, based on key indicators of activity (KIA).

The content statement serves as a basis for forming an optimization model for selecting the KIA for the deviation management problem.

Invariant descriptions of professional activity models have been developed: an inductive logical model for assessing the reasons for the deviation of the actual KIA values from the planned ones;

optimization model of management decision-making to compensate for deviations.

Chapter 5 is devoted to solving the problem of optimizing the training of personnel for civil aviation on the basis of professional activity models in the development of airport infrastructure systems. Models that show promising competencies in the development of airport infrastructure systems should be focused on two main classes:

geographically distributed airport systems;
internal facilities and airport complexes.

The most acceptable form of invariant description for the first class is situational models that allow us to assess the initial state of the airport infrastructure of the central hub and regional airport infrastructure and consider the possibility of moving to a new state with better characteristics by making certain decisions on the development of these complex systems.

Among the internal facilities and airport infrastructure complexes, the leading requirements are primarily applied to radio communication systems, radio navigation and information systems. Therefore, it is necessary to distinguish professional activity models for the following components:

radio communication systems as an integral part of an automated air traffic control system;
intra-airport communication;
radio navigation systems;
aviation electrical systems;
airport information systems.

The scheme of multi-stage reduction and subsequent transformation of professional activity models into components of educational programs is used.

Chapter 6 presents the results of the development of innovative components of the modernization of the training process and assesses their effectiveness in training personnel in the field of air transport operation. It is shown that the innovative components that implement the model-based approach methodology in the system of civil aviation universities primarily concern the modernization of the structure of educational organizations and the content of training based on new generation standards.

To solve the first problem, a structural and functional model of an educational organization in the form of hierarchical relations of objects of a branch educational cluster is proposed. Taking into account that the elements of these relations are in an ambiguous interaction, leading to uncertainty of the system properties, the necessity of algorithmization of the integrated assessment of the quality of the educational system, based on the methods of expert monitoring analysis, is shown. On the basis of a comparative analysis of integrated assessments for the modes of operation and modernization, decisions are made on a set of innovative measures.

The solution to the problem of optimal sequence of training is based on a hierarchical model in the form of a tree of goals, the transitions between the levels of which correspond to the decomposition of educational standards of the new generation. A distinctive feature of the application of the model-oriented approach is the description of the elements of the lower levels on the basis of a reduced set of models of professional activity, which at the last level are transformed into components of the educational resources of the upper levels.

The results of the analysis of the effectiveness of the model-based approach in the modernization of the educational cluster of civil aviation are presented.

Contents

Abbreviations

AIP	Automated information processing
ASC	Automated system of control
ATIS	Automatic transfer of information in the air station areas
ATM	Air traffic management
CA	Civil aviation
CFA	Control figures of admission
EU	European Union
FSEPS	Federal State Educational Professional Standards
GIS	Geographic information system
HE	Higher education
HEI	Higher Educational Institute
IS	Information system
KAS	Knowledge, abilities, skills
KIA	Key indicators of activity
KSA	Knowledge, skills, abilities
LSM	Least squares method
MAH	Moscow Aviation Hub
MSTUCA	Moscow State Technical University of Civil Aviation
OFIS	Operational flight information service
RDW	Research and development work
RF	Russian Federation
RW	Research work
SPE	Scientific and pedagogical employee
SPSUCA	St. Petersburg State University of Civil Aviation
SPW	Scientific and pedagogical worker
SVE	Secondary vocational education
UI CA	Ulyanovsk Institute of Civil Aviation
USE	Unified State Examination

Chapter 1
System Analysis of Problems of Personnel Training Management With Professional and Higher Education in the Field of Air Transport Operation

An essential component of theoretical and applied research, related to improving the efficiency of the air transport operation process, is the development of mechanisms and tools for improving the personnel training system [1–8]. However, such studies are more concerned with local problems of personnel support for civil aviation, aimed at:

- improving the level of flight safety [9–12];
- improving the professional psychophysiological selection of aviation specialists [13, 14];
- improving models and means of organizing training in certain types of professional training [15–19];
- modernization of aviation personnel training in certain educational organizations of the industry [20].

Attempts to consider the domestic experience of organizing the training of aviation specialists from a systematic perspective are more concerned with the historical and pedagogical aspect [21]. Unfortunately, in the works, devoted to the assessment of the development of professional education in civil aviation, taking into account the scientific and technological progress and modernization of the Russian economy, most often, these tasks are considered only in the format of problem formulation [22].

Therefore, the development of a unified methodological framework for optimizing personnel training in the field of air transport operation requires a systematic analysis of the main set of problems and the links between them in the chain "industry educational cluster—aviation enterprises" with a focus on the requirements of employers for the formation of the main competencies of specialists that meet professional standards for the modernization of the training system, forecasting the need for personnel for all types of air transport operation, taking into account the requirements of employers.

A. Borzova et al., *Modeling and Forecasting of Staffing in Civil Aviation*,
Springer Aerospace Technology, https://doi.org/10.1007/978-981-19-6238-7_1

1.1 Ensuring the Impact of Innovative Approaches to the Operation of Air Transport System Requirements Training

The implementation of innovative approaches in the operation of air transport is due to several factors, one of which is the role and importance of air transportation in the country's economy. Meeting the need for passenger and cargo transportation with a high level of safety, regularity and quality of service is an essential condition for the socioeconomic life of the regions of Russia.

The volume characteristics of transport services achieved in accordance with the implementation of the "Transport Strategy of the Russian Federation until 2030", approved by the decree of the Government of the Russian Federation No. 1734-r of November 22, 2008, as well as the targets of the State Program "Development of the Transport System for 2016–2021", approved by the decree of the Government of the Russian Federation of April 15, 2014, show the results of the economic development of the country, demonstrating its competitiveness at the world level (Figs. 1.1, 1.2 and 1.3).

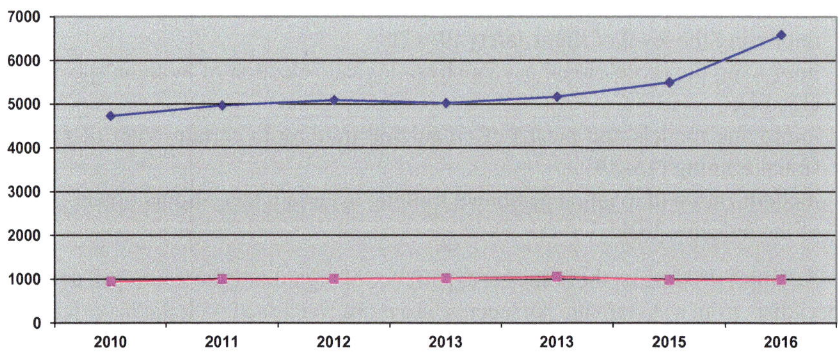

Fig. 1.1 Cargo turnover, transportation of cargo and mail by air transport

Fig. 1.2 Passenger traffic and passenger transportation

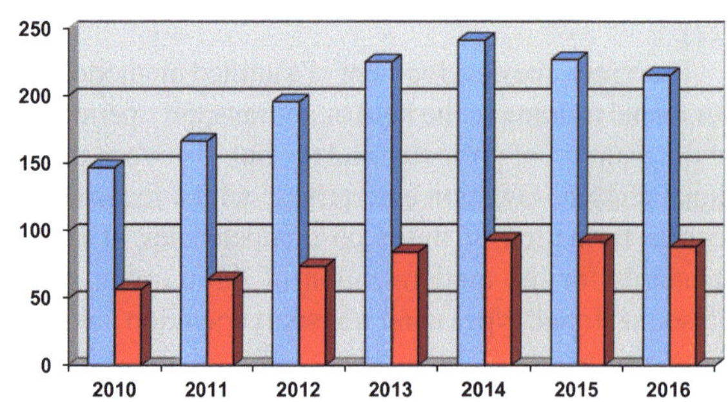

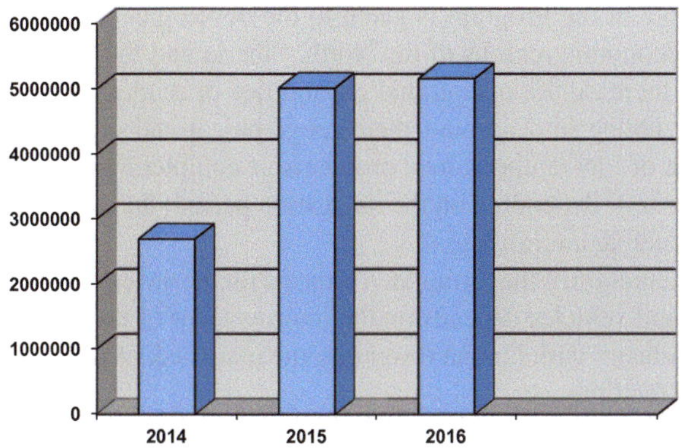

Fig. 1.3 Passengers, served at Simferopol Airport

Cargo turnover in 2016 amounted to 6575 million tons km (an increase of 39.45% compared to 2010—4715). Cargo and mail transportation in 2016 amounted to 977 thousand tons (an increase of 5.5% compared to 2010—926).

In Fig. 1.1, the blue line shows the cargo turnover in million tons km, and the purple line shows—transportation of cargo and mail in thousand tons.

In Fig. 1.2, the blue column shows passenger traffic in billions passengers km, and the red column shows passenger traffic in millions of passengers.

Passenger traffic in 2016 amounted to 215.59 million passenger kilometers (an increase of 46.56% compared to 2010—147.1). Passenger traffic in 2016 amounted to 88.56 million passengers. (An increase of 55.5% compared to 2010—56.95).

Analyzing the data, presented in Fig. 1.3, it is established that the increase in the volume of air traffic was achieved by increasing the attractiveness of the Republic of Crimea and the resorts of the Krasnodar Territory in 2016. In this regard, the number of passengers served at Simferopol airport was more than 90%. The popularity of this region, especially in the summer period, provided a significant increase in passengers at the airports of the Krasnodar Territory, by almost 26% in 2016 compared to 2015.

It is proved that the achievement of a high level of volume indicators of both individual airlines and the industry as a whole depends on the demand for transportation. One of the factors in the formation of demand for air transportation is the quality of transport services, including speed of delivery; regularity of transportation, safety and environmental friendliness of air transport infrastructure facilities.

The book analyzes the Program for the Development of the Air Transport Industry until 2030, which sets out the primary objectives for the development of leading civil aviation infrastructure facilities. It is established that the main focus of the Program is focused on the following tasks:

- modernization of the ground infrastructure of air transport, consisting of airfields of international and domestic hub airports;
- development of infrastructure for the development of business transportation;
- improving the efficiency of the air traffic management system;
- improvement of the meteorological support system.

A special role in the Program is given to the development of air transportation in the remote economic regions of the North, Siberia and the Far East. This is due to the need to increase the operational capabilities of modern aircraft for regional transportation, taking into account their geographical and climatic features, such as the presence of low temperatures; problems of completing the range of aviation fuels and lubricants, depending on the navigation period; the creation of centralized systems for refueling aircraft, etc.

The implementation of the formulated projects for the development of air transport infrastructure and vehicles depends on the improvement of the system of providing the transport industry with human resources, the main task of which is the operation of modernized facilities.

Unfortunately, the presented Program for the Development of the Air Transport Industry until 2030 does not sufficiently show the tasks of training air personnel, on the success of which the achievement of the set goals can significantly depend.

This fact determines the need to develop new formats of educational organizations that implement training of specialists in the field of civil aviation, innovative educational programs and updating existing ones in accordance with the modern requirements of the dynamically developing labor market, the implementation of the principle of training specialists "during all life", based on the continuity of various levels of professional education (secondary, higher, additional and postgraduate), the formation of individual training trajectories, integration and implementation of mobile and electronic technologies in the educational process.

1.2 Achieving Compliance with the Characteristics of the System of Training Needs of Enterprises

The results of numerous scientific studies emphasize the need to create an effective system of training specialists that meet the needs of enterprises in all areas of activity. For the air transport system, which provides special conditions for economic growth and competitiveness of the country, this task is important, since the implementation of a set of interrelated measures for the modernization of civil aviation in the future depends on the effectiveness of the mechanism for managing labor resources and the formation of personnel policy in the transport complex.

The analysis of the existing system of training, retraining and advanced training of civil aviation specialists and managers allowed us to identify a number of shortcomings, contradictions and discrepancies that reduce the effectiveness of training, depending on the dynamic processes and changes taking place in the air transport industry. One of these disadvantages is the low level of relationship between theoretical training and the formation of practical skills and abilities.

Another drawback in the training of specialists for civil aviation is the use of a competence-based approach, which does not fully take into account the personal qualities of the trainee in modern conditions.

In addition, there are no clear procedures for improving the quality of training of aviation specialists that meet the requirements of international standards and practices.

In this regard, there is a need to create a new system of continuous professional training and retraining of aviation personnel. A special feature of such a system is the development and scientific justification of a multi-level socio-educational model that meets the needs of the aviation market, industry, educational institution and individual.

Based on this, the monograph examines the issues of a scientifically based approach to the process of training aviation personnel, which include several main areas.

First is the development and scientific justification of the methodological framework and the mechanism for monitoring and regulating personnel support in the transport industry.

Second is the creation of an adapted system for the formation of a reserve of professionally trained personnel in the field of air traffic control, aircraft maintenance and flight safety.

Third is the formation of a system of training aviation professionals, who have the skills to maintain the latest aviation and ground equipment.

Fourth is conducting scientific research and developing measures that increase the prestige of aviation professions, creating a high level of competitiveness of civil aviation educational institutions.

For the practical implementation of the tasks set, the book formulates approaches to the formation of a continuous multi-level educational system built on the basis of vertical integration of aviation educational institutions of various types and levels (Fig. 1.4).

According to the presented scheme, increasing the role of specialized aviation training centers is achieved through the formation of vertically integrated links

Fig. 1.4 Structure of the civil aviation training system

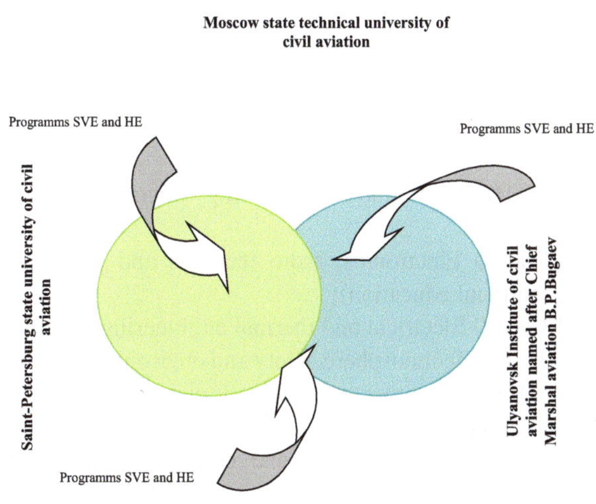

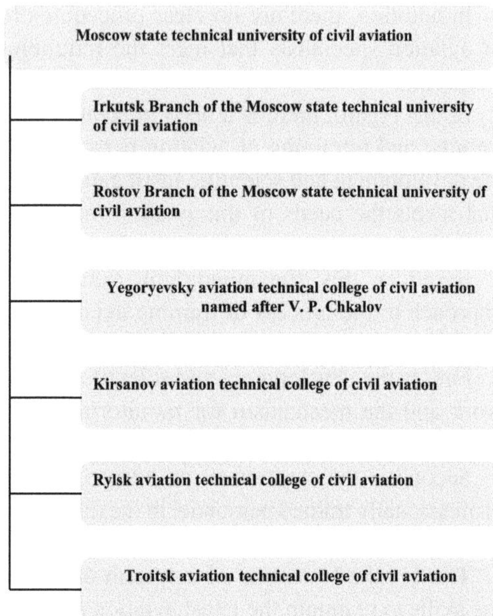

Fig. 1.5 Moscow State Technical University of Civil Aviation and its branches

in the socioeconomic, technological and educational development of the aviation industry in various regions of Russia: the international hub airports of the Moscow aviation hub (Domodedovo, Vnukovo, Sheremetyevo); Yekaterinburg, Novosibirsk, Khabarovsk, Krasnoyarsk, Samara, St. Petersburg, Kaliningrad, Volgograd, Omsk, Blagoveshchensk, Nizhny Novgorod, Ufa, Perm, Chelyabinsk, Sochi, Anapa, Mineralnye Vody, Astrakhan, Penza, Saratov, Nizhnevartovsk, Barnaul, Magnitogorsk, Kemerovo, Novokuznetsk, Bratsk, Voronezh, Vorkuta, Khanty-Mansiysk and others.

Educational institutions of civil aviation of the Russian Federation have separate structural divisions (branches) that provide training for specialists in both higher and secondary vocational education programs (SVE и HE) (Figs. 1.5, 1.6 and 1.7).

The educational process is a continuous vertically integrated training for eight large groups of directions (specialties), taking into account the profile of training to ensure the functioning of civil aviation:

- 08.00.00 Technique and technology of construction (secondary vocational education);
- 11.00.00 Electronics, radio technics and communication systems (secondary vocational education);
- 13.00.00 Electrical and thermal engineering (secondary vocational education);
- 20.00.00 Technosphere safety and environmental engineering (higher education);
- 23.00.00 Equipment and technologies of ground transport (secondary vocational and higher education);
- 25.00.00 Air navigation and maintenance of aviation and rocket and space technology (secondary vocational and higher education);

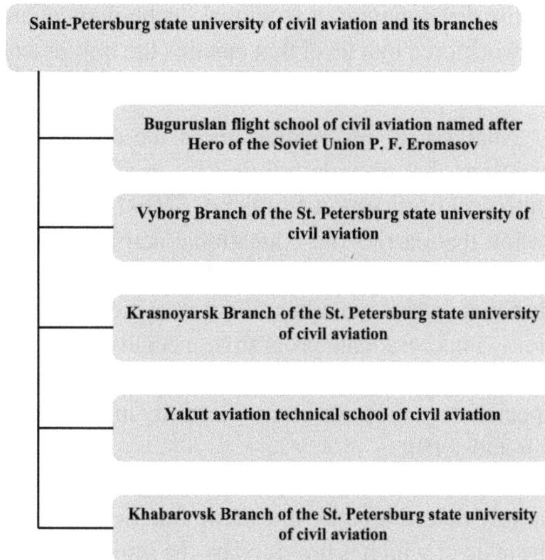

Fig. 1.6 Saint-Petersburg State University of Civil Aviation and its branches

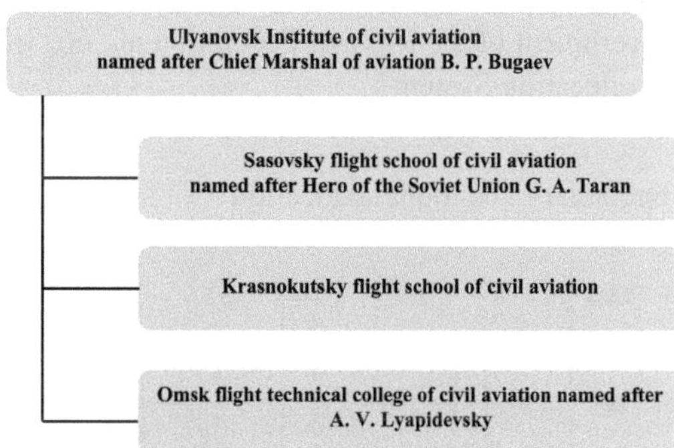

Fig. 1.7 Ulyanovsk Institute of Civil Aviation named after Chief Marshal of Aviation B. P. Bugaev and its branches

- 27.00.00 Management in technical systems (higher education);
- 43.00.00 Service and tourism (secondary vocational education).

Such a training structure provides the necessary structure for training specialists and labor resources for civil aviation (transport system) in the following areas:

development of human resources in the field of design and implementation of projects for the development of the transport system;
development of the provision of labor resources in the field of operation of transport infrastructure and vehicles;
development of human resources provision in the field of transport and logistics services and other transport services;
development of human resources provision in the field of transport complex management;

the development of technical, technological and other types of knowledge of the workforce to a level that ensures the implementation of the goals of the transport strategy.

Analysis of information about the contingent of students in educational organizations that provide training for civil aviation industries formed on the basis of statistical observation forms no. SVE-1 "Information about an educational organization that carries out educational activities in educational programs of secondary vocational education" and no. HVE-1 "Information about an educational organization that carries out educational activities in educational programs of higher education—undergraduate programs, specialty programs, master's programs", showed the dynamics of increasing the intake, and, accordingly, the contingent and graduation of specialists in the main areas of activity in the field of air transport operation (Figs. 1.8, 1.9 and 1.10).

The effectiveness of personnel support in the field of air transport operation is largely determined by the stable functioning of the training system, the analysis of the state of which is provided by the implementation of the monitoring procedure for the dynamics of system indicators, which is carried out annually by the Ministry of Education and Science of the Russian Federation in accordance with the Decree of the Government of the Russian Federation no. 662 of August 5, 2013, "On monitoring the education System".

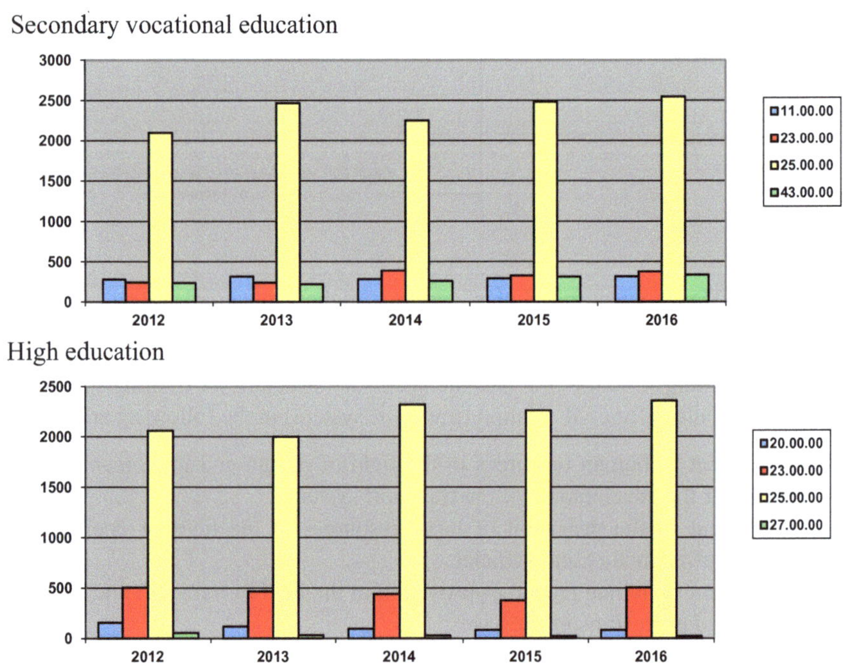

Fig. 1.8 Admission to educational organizations of civil aviation

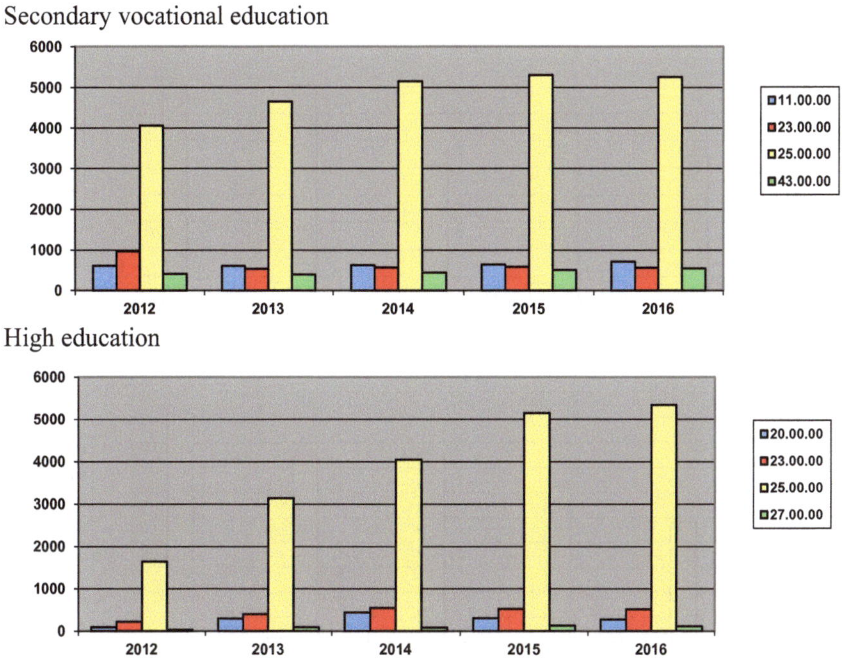

Fig. 1.9 Number of students in educational organizations of civil aviation

The purpose of evaluating the effectiveness of professional educational organizations is to increase the effectiveness of their activities, aimed at improving the educational process, expanding its integration with scientific activities, developing international cooperation with foreign partners and expanding competitiveness.

The evaluation of the effectiveness of professional educational organizations is based on the analysis of the results of their activities in comparison with the criteria of the all-Russian evaluation system. The use of certain criteria of the all-Russian system for evaluating the effectiveness of educational institutions depends on the objects of monitoring the state of the vocational education system. In this regard, it is necessary to distinguish the following objects: educational activities; scientific and innovative activities; organizational and structural activities; commercial activities; personnel support; financial and logistical support; socioeconomic support.

Thus, the results of the comparative analysis of educational organizations of higher education in the field of transport for 2014–2016 are presented as follows:

1. Educational activities: The indicator characterizes the average score of the

   Unified State Exam (USE) of students, enrolled in full-time bachelor's and specialist's degree programs (Fig. 1.11). The light blue column characterizes rail transport, the red column characterizes sea and river transport, the yellow column characterizes road transport, and the dark blue column characterizes air transport.

Secondary vocational education

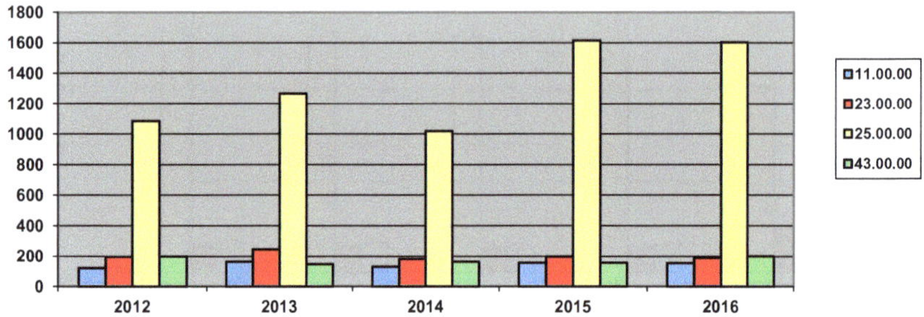

High education

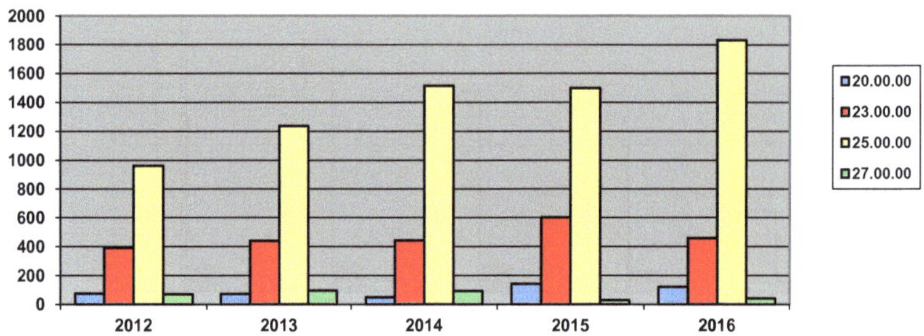

Fig. 1.10 Graduation of specialists in the field of air transport maintenance

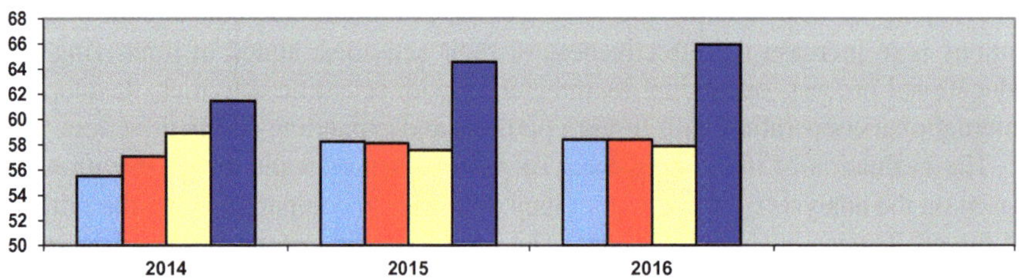

Fig. 1.11 Dynamics of the effectiveness of educational activities

2. Research activities: The indicator characterizes the volume of research and development work (RDW) per scientific and pedagogical employee (SPE) (Fig. 1.12).
3. International activities: The indicator characterizes the share of the number of foreign students in bachelor's, specialist's and master's programs in the total number of students (Fig. 1.13).
4. Financial and economic activities: The indicator characterizes the income from all sources of the educational organization per one scientific and pedagogical employee (SPE) (Fig. 1.14).

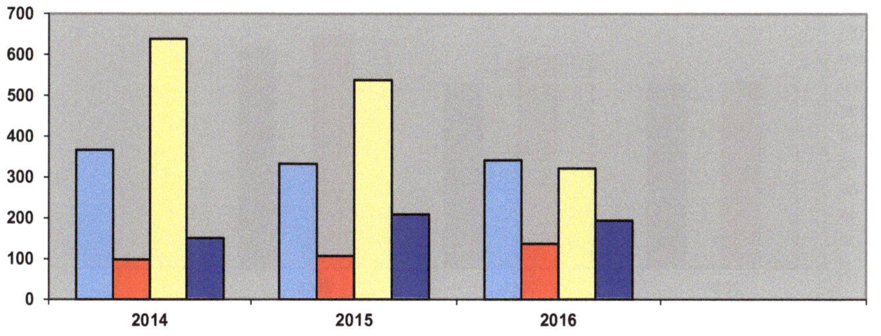

Fig. 1.12 Dynamics of the effectiveness of research activities

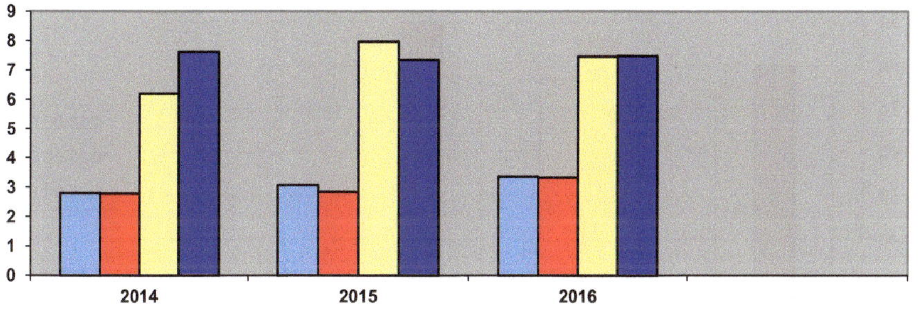

Fig. 1.13 Dynamics of the effectiveness of international activities

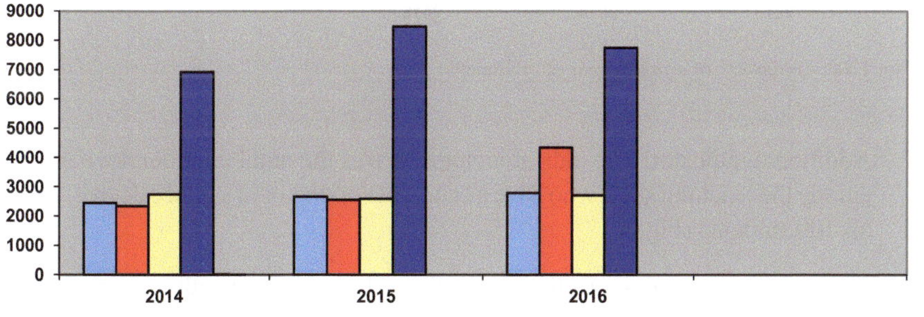

Fig. 1.14 Dynamics of the efficiency of financial and economic activities

5. Salary of scientific and pedagogical employees: The indicator characterizes the ratio of the average salary of the teaching staff to the average salary in the economy of the region (Fig. 1.15).
6. Employment: The indicator characterizes the share of graduates, employed during the calendar year, following the year of graduation in the total number of graduates of an educational organization in higher education programs (Fig. 1.16).

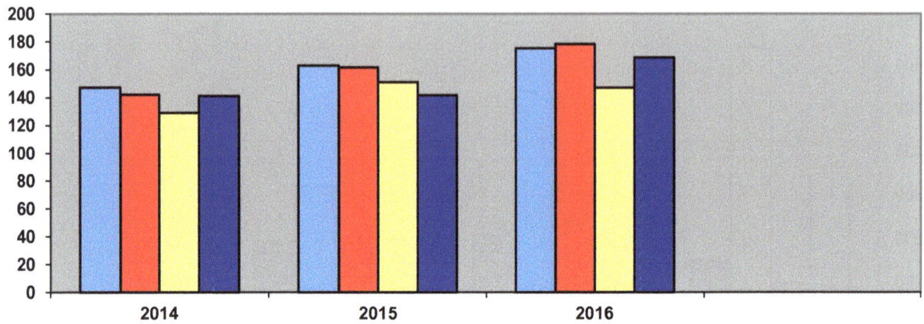

Fig. 1.15 Dynamics of SPE's salary

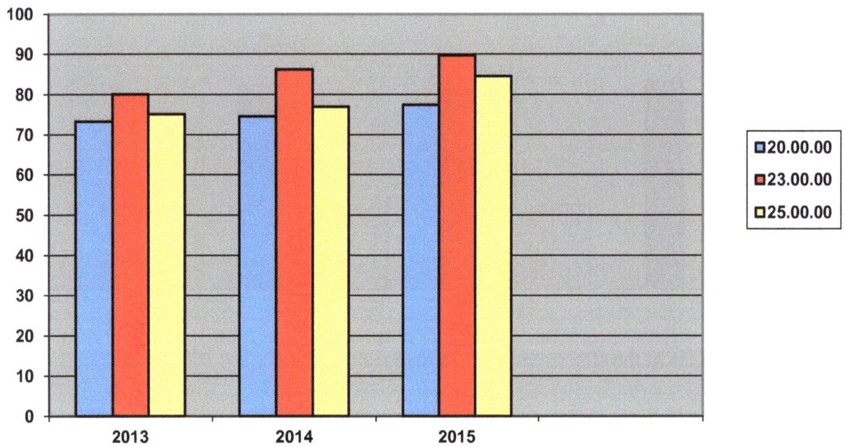

Fig. 1.16 Dynamics of employment of graduates

7. Additional indicator: The indicator characterizes the number of employees from among the teaching staff, who have a candidate or doctor of science degree, based on 100 students (Fig. 1.17).

Analysis of the dynamics of changes in the values of the results of monitoring the activities of educational organizations that provide training for civil aviation specialists allows us to achieve the following goals:

- ensuring the sustainability of the functioning and development of the system of training specialists 1.18;
- forecasting the development of the educational system as a whole;
- formation of policy in the field of development of the system of training of specialists.

To achieve these goals in the format of the activities of educational institutions of civil aviation, it is necessary to solve the following tasks:

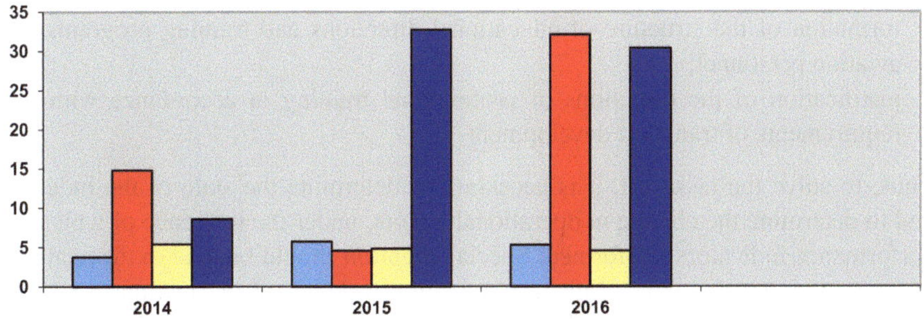

Fig. 1.17 Dynamics of personnel support efficiency

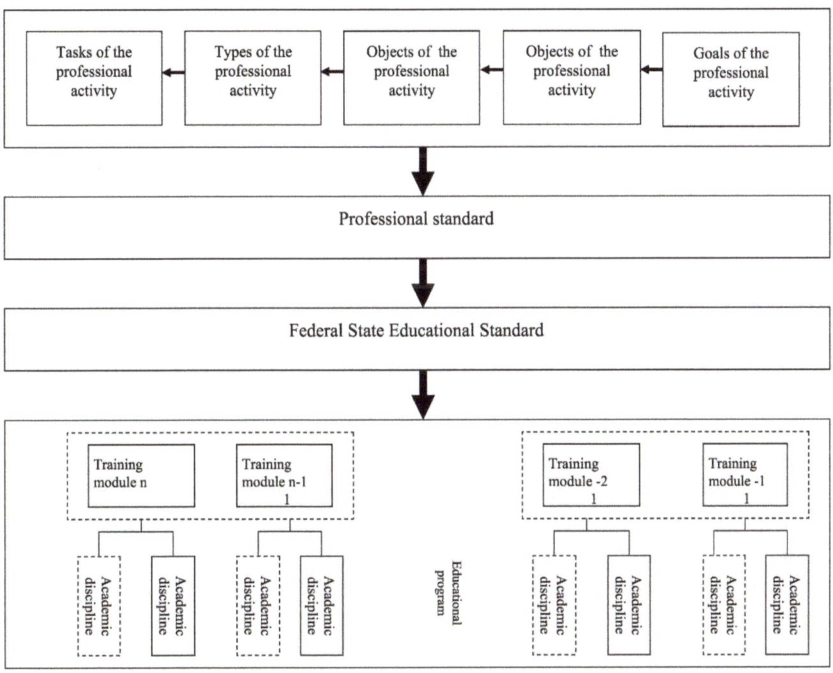

Fig. 1.18 Scheme of the practice-oriented approach to develop educational programs

- study and analysis of the general state of the current educational system;
- determination of the reasons–factors, influencing the effectiveness of the educational system;
- substantiation of the quality parameters and determination of the degree of their compliance with the required number of aviation specialists in the industry and the structure of training;
- justification of the mechanism for monitoring and managing the system of educational institutions of regional significance;

- formation of the structure of educational directions and training programs for aviation personnel;
- justification of the directions of professional training in accordance with the requirements of transport development.

Thus, to solve the tasks set, it is necessary to determine the state of the industry and to determine the change in operational factors, under the influence of which its performance indicators are formed. Special attention should be paid to the analysis of innovative development of air transport, improvement of production processes and management organization. This will make it possible to make a forecast of the personnel training system, determine the content of future education and adapt the personnel training system to the needs of the personnel supply industry.

1.3 The Need for the Convergence of Competence Characteristics Educational and Professional Standards

Modern education is focused on the labor market. This lies in the requirements, imposed on educational organizations of higher and secondary vocational education and on their graduates.

The professional standard as a document that establishes the qualification level of an employee, allowing him to perform his official (professional) duties in accordance with the requirements for a specific position (profession), is the foundation for the development of educational standards, the graduate's competence model and the main professional training programs [23–27]. At the same time, the educational standard is a set of knowledge and skills that a student must acquire in an educational organization in order to further perform their official (professional) duties.

For professional educational organizations, professional standards make it possible to build activities in "real time" and to determine the trajectory of practice-oriented training, based on the consolidation of education, science and industry employers.

In this regard, the results of the analysis of the implementation of the requirements of the Federal State Educational Professional Standards (FSES) in terms of labor functional responsibilities, knowledge, skills and abilities are of the greatest interest.

The algorithm for analyzing the implementation of the requirements of the Federal State Educational Standard, corresponding to a specific field of training, can be formulated in several stages.

Stage 1 Study of the features of the professional profile of a particular field of activity;

Stage 2 Formulation of a list of knowledge, skills and practical skills for performing specific professional functions;

Stage 3 Development of educational training modules that correspond to the development of professional competencies;
Stage 4 Formation of a list of additional professional competencies to expand the structure of professional functions;
Stage 5 Ranking of professional competencies by the degree of their development.

A comparison of one of the main activities of a specialist in the field of air transport operation, the labor functions, defined by the professional standard "Specialist in after-sales service of aviation equipment" (Order of the Ministry of labor of the Russian Federation of 21.12.2015 no. 1052n), as well as professional competencies, defined by the Federal State Educational Standard for the enlarged group of areas (specialties) of training 25.00.00 "Air navigation and maintenance of aviation and rocket and space equipment" is presented in Table 1.1 [28–41].

Thus, the comparison of labor functions in professional standards and competencies in educational standards allows to perform their integration within the framework of knowledge, skills and abilities, provided by educational resources, formed in the studied disciplines of educational programs of higher and secondary vocational education.

1.4 Ways to Improve the Efficiency of Personnel Training Management at Aviation Enterprises, Based on Modeling and Optimization Methods

The process of functioning and development of the air transport maintenance industry is determined by the influence of many factors of social, economic, technological and other processes and the time factor accumulates their influence.

In particular, the analysis of ways to improve the efficiency of management of personnel training of aviation enterprises can be carried out on the basis of the use of forecasting, modeling, decision-making and optimization methods [42–49].

In order to develop a modern management system for training personnel of aviation enterprises, the following stages of the modeling process are used [50–54]:

description of the object, problem statement and definition of goals and objectives;
building a basic model that describes the subject of research by applying a set of absolute statistical, relative structural and qualitative indicators that allow you to choose the methods and methods of research;
collection of statistical data on the simulated system;
diagnostics of the evolutionary development of the system in order to identify patterns, trends, contradictions and shortcomings. Identification of obvious and hidden opportunities for the development of a training system for the air transport industry to assess the necessary resources and sources of their formation;
building a normative and/or search model of the system under study;
verification of research results.

Table 1.1 Example comparison of the type of activity of a specialist in the field of maintenance (training direction 25.00.00) of air transport, labor functions and professional competencies

Type of activity	Labor functions	Professional competencies of FSES
Research and development, aimed at improving the airworthiness of aircraft, improving maintenance and technical characteristics, changing the properties of technical objects, improving the efficiency of the technical maintenance system, improving the regulatory and technical documentation and information base	Material and technical support of the aviation equipment maintenance process	Ability to conduct research to reduce the loss of material resources, labor and time in the process of technical operation of aircraft; Aircraft operation and maintenance readiness; The ability and readiness to operate aircraft, power plants and aircraft systems, including radio and electric lighting equipment, automation and control systems and on-board emergency rescue equipment, in accordance with the requirements of regulatory and technical documents, the requirements of air legislation, federal aviation regulations and regulatory legal acts of the Russian Federation; Ability to manage the technical condition of aircraft equipment, the efficiency of production processes at the stages of operation of aircraft equipment; The ability to develop standards, production and technological standards for the consumption of materials, fuel and electricity, as well as to justify the choice of equipment and technological equipment, algorithms and programs for calculating the parameters of the technological process; Ability and willingness to select replacement equipment during the operation of aircraft and aviation infrastructure facilities; Ability to configure and maintain hardware and software; Ability and readiness to check the technical condition and residual life of the equipment, to organize and provide preventive inspections and routine maintenance of the equipment; Ability and willingness to accept and master the equipment being introduced; ability to make requests for equipment, materials and spare parts; Ability and willingness to organize, execute, provide and maintain aircraft operations; Ability and willingness to organize and implement air traffic services (management); Ability and willingness to organize, perform, provide and maintain air transportation and aviation operations; The ability and readiness to organize, provide and carry out measures to ensure the safety of aircraft flights and the safety of the use of airspace, to ensure aviation security, to ensure the quality of work and services; The ability and willingness to carry out aircraft flight planning, create operational flight plans and flight plans for the purposes of air traffic services

(continued)

Table 1.1 (continued)

Type of activity	Labor functions	Professional competencies of FSES
	Ensuring the quality of after-sales service of aircraft equipment	The ability and willingness to check the operability of the equipment in use; Ability to apply new modern methods of development of technological processes of maintenance, repair of aircraft equipment with the definition of rational technological modes of equipment operation; Ability to manage (calculate) the required resources to ensure the process of maintaining the airworthiness of aircraft, including production facilities, personnel, equipment, tools; Ability to address issues of quality assurance of maintenance and repair of aircraft outside the base (airline); The ability to participate in carrying out a complex of planned preventive works to ensure the serviceability, operability and readiness of aircraft facilities for effective use for their intended purpose; Readiness to carry out verification of the technical condition and residual life of aircraft equipment and equipment, to organize preventive inspections and routine repairs in order to maintain the airworthiness of aircraft and ensure flight safety; Knowledge of the organizational structure, management methods and regulation of performance criteria in relation to specific types of aircraft operation, storage, refueling, maintenance and repair of aircraft equipment; Ability to organize and conduct quality control of aircraft maintenance and repair, compliance with state requirements for maintaining airworthiness and ensuring flight safety during the maintenance of aircraft equipment; Knowledge of the system of maintenance and repair of aircraft equipment and technological equipment, methods of ensuring the safety of operation (including environmental), storage, maintenance of aircraft equipment and equipment, safe working conditions of personnel; Ability to assess the technical and economic efficiency of aircraft maintenance and technological processes, willingness to participate in the development of recommendations for improving the operational and technical characteristics of aircraft operation

(continued)

Table 1.1 (continued)

Type of activity	Labor functions	Professional competencies of FSES
	Information and regulatory support for the operation of aviation equipment	Ability to study objects and processes of aviation equipment operation and to analyze the obtained results, including with the help of application software packages and elements of mathematical modeling; The ability and readiness to organize and perform work on information support for the operation of aircraft and aviation infrastructure facilities, air traffic management, air navigation services for aircraft flights and the use of airspace, using computer technology; The ability to solve the problems of planning the technical operation of aircraft, operational reliability, flight regularity, as well as the organization, information and hardware support of the production processes of aircraft maintenance and repair and cost-effectiveness of use; The ability to compile and maintain technical documentation and established reporting in accordance with approved forms, including accounting for the resource and technical condition of aircraft, as well as ensuring the regulatory working conditions of employees of the engineering and aviation service, fire safety and environmental protection; The ability to make requests for the necessary technical equipment and spare parts, to prepare technical documentation for repairs; The ability to develop operational documentation that regulates the maintenance of aircraft and aviation infrastructure facilities
	Organization of the provision of data for the calculation of the economic efficiency of the maintenance of aviation equipment	The ability and readiness to prepare data for decision-making in the management of transport systems in various conditions, to analyze the effectiveness of the functioning of transport systems; The ability to conduct contractual work on production activities and prepare initial data for the selection and justification of scientific, technical and organizational decisions, based on economic analysis; Ability and willingness to use micro-and macroeconomics methods in solving professional tasks; The ability and willingness to determine the financial results of the production and economic activities of aviation enterprises; ability and willingness to calculate and evaluate the conditions and consequences (including economic ones) of organizational and managerial decisions made; The ability and willingness to organize and conduct a marketing analysis of the need for services in the maintenance of aircraft and aviation infrastructure facilities, in the organization, execution, provision and maintenance of aircraft flights, air transportation and aviation operations

Forecasting methods are an integral part of managerial decision-making in the management system of personnel training for airlines (Fig. 1.19).

Statistical methods are aimed at predicting the results of the implementation of the adopted strategies. The assessment of uncertainty in the implementation of the strategic objectives of the training system can be obtained through the use of expert assessments. With a limited source of information and the ability to obtain accurate estimates of development, the forecast can be made, based on the method of analogies with feedback.

Figure 1.20 shows a diagram of the stages of forecasting in aviation enterprises.

The main task of forecasts is the application of an integrated approach, including forecasting the number and structural composition of personnel, depending on the forecasting of volume indicators of activity, the emergence of new and the use of current resources of the enterprise, forecasting changes in economic indicators of activity.

One of the methods of evaluating the effectiveness of the functioning and development of the system of training personnel in the field of air transport maintenance, aimed at ensuring the activities of Russian civil aviation enterprises, can be used statistical methods of analysis, in particular, time series forecasting [55]. The essence of this method is that based on the visual study of the graph and its analysis, deviations of the time series parameters are determined. According to these parameters, the level of deviation of the empirical values from the theoretical values is determined, and, thus, an equation is selected for describing and constructing the trend line.

The detected deviations can be absolute and relative. The value of absolute deviations indicates a normal additive distribution between the components and in the presence of relative deviations—multiplicative.

Depending on the effect of the components under consideration on the value of the time series, the following models can be used.

For example, if multiplicative factors occur, the values of the time series levels are described by the following model

$$\overline{y_t} = T \cdot C \cdot \varepsilon; \tag{1.1}$$

- if the factors are additive, then the values of the time series levels are

$$\overline{y_t} = T + C + \varepsilon; \tag{1.2}$$

- if the time series factors are expressed in combination, then the values of the time series levels are

$$\overline{y_t} = T \cdot C + \varepsilon, \tag{1.3}$$

where T is trend (a general long-term pattern in changing the values of a time series over a long period of time);

C is seasonal component (repeated in relation to the trend of movement up and down, the cycle of which is usually completed at the end of the year, and in subsequent years the cycles are repeated again); ε is random component.

The influence of the seasonal component is manifested depending on the behavior of the absolute deviation. For example, absolute deviations increase dynamically, while relative deviations stagnate or vary at the same level. This state of the system indicates the presence of a multiplicative relationship between the trend and the seasonal component.

If we assume that probabilistic estimates are used in the assessment of ongoing processes in the education system, then the levels of the time series can be represented as a sum that characterizes the change in deterministic and random processes.

The generalized indicators of the characteristics of the dynamics series in the analysis of changes in the system of vocational education include the average indicators, for example the average level of the series in the study period. The essence of its value is due to the identification of stability in the processes, depending on the occurrence of force majeure circumstances that cause random fluctuations in the levels of the series.

$$\overline{y} = \frac{\sum_{t=1}^{n} y_i}{n}, \tag{1.4}$$

where n—the number of levels or the length of the row; y_t—the level of the time series.

According to the structure of the existing forecasting methods (Fig. 1.19), expert methods based on the methods of direct expert assessment, expert survey and expert analysis are used. The system of expert evaluation includes a database of objects subject to examination; a database of experts [56, 57].

The role and importance of expert assessments are determined by the justification of optimal models and the choice of the most preferred option for the formation of the content part of training programs for specialists in the field of air transport maintenance. As a tool for selecting the preferred option, a method adapted to the conditions under consideration can be used. Suppose that there is some arbitrary set of parameters $A = \{a_1, \ldots, a_\mathrm{n}\}$ that characterize the features of a specific program of training a specialist for civil aviation (e.g., the level of complexity, training time, the minimum number of practical tasks, etc.) and a set of criteria $K = \{k_1, \ldots, k_p\}$, for evaluating qualitative (e.g., the level of development of a specific program), or quantitative assessments of alternative options (e.g., the absolute number of evaluation points) [58].

When conducting a pairwise comparison of alternative options that characterize the features of the training program and the degree of its development for an arbitrary pair of alternatives a_i, $a_i \in A$, a situation may arise: $a_i > a_j$; $a_j > a_i$; $a_j \approx a_i$; a_i and $\mathrm{a_j}$ are incomparable. In this case, you should use the following method. If the received scores are based on the criteria $K_1, \ldots, K_m$ and quantitative, then the

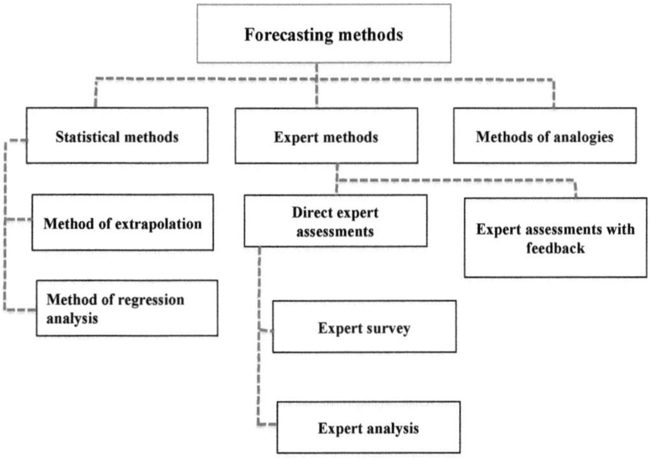

Fig. 1.19 Forecasting methods in management decision-making

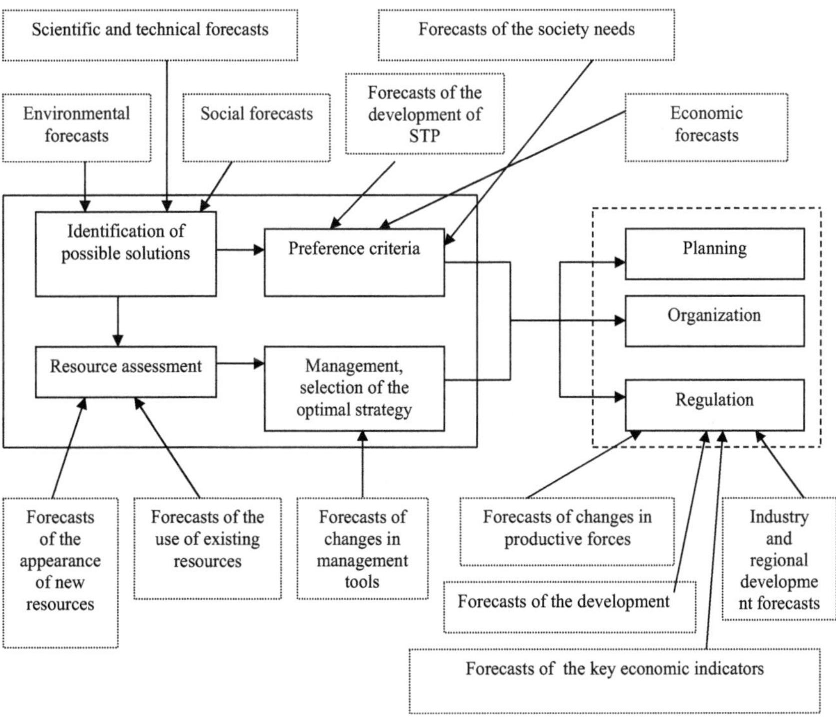

Fig. 1.20 Stages of forecasting in decision-making

choice of an alternative option is based on a comparison of the evaluation vectors $X_1 = (x_1^i, ..., X_m^i)$ alternative a_i and $X_j = (x_1^j, ..., X_m^j)$ alternative a_j.

If the obtained estimates according to the applied selection criteria are of a qualitative nature, then the conclusion on the comparative preference of alternatives a_i and a_j is made, using comparison vectors $S_{ij} = (s_1^{ij}, ..., s_m^{ij})$, where

$$S_t^{ij} = \begin{cases} 1,\ if\ a_i > a_j\ by\ criteria\ K_t,\ t = \overline{1, m} \\ 0, if\ a_i \approx a_j\ by\ criteria\ K_t,\ t = \overline{1, m} \\ -1,\ if\ a_i < a_j\ by\ criteria\ K_t,\ t = \overline{1, m} \end{cases} \tag{1.5}$$

If the evaluation of alternatives a_i and a_j by all criteria is quantitative, then it is advisable to consider metrized comparison vectors $S_{ij} = (s_1^{ij}, ..., s_m^{ij})$, , where (1.6)

$$S_t^{ij} = X_t^i - X_t^j, \quad t = 1, m \tag{1.6}$$

In this case, the choice principle π determines the choice on the pair of alternatives a_i and a_j, if according to the principle of choice π $a_i > a_j$, $a_j \approx a_i$ or $a_j > a_i$. Otherwise, π does not determine the choice on the pair of alternatives a_i and a_j. The set of comparison vectors S_{ij} that determine the choice according to the principle of choice π is denoted by S^0.

Any principle of choice can be assigned to one of four types:

π_1—fully defines a choice, characterized by a non-metrized comparison vector (1.5);

π_2—fully defines a choice, characterized by a metrized comparison vector (1.2);

π_3—fully defines a choice that is not characterized by comparison vectors;

π_4—does not fully define a choice.

Another method for a choice of alternatives may be a choice by determining the equivalence relation $P(k)$ on a set of criteria $K = \{\kappa_1, \ldots, \kappa_m\}$ [59, 60]. The essence of the method is that the set of criteria $\kappa_1, \ldots, \kappa_m$ is divided into classes that are homogeneous in their content. Next, criteria are selected that form equivalence relations $P_f(k)$ based on the preference of individual criteria, i.e., $a_i > a_j$,, if $K_1(a_i) > K_1(a_j)$, $K_2(a_i) > K_2(a_j)$ and any value

$$c = \frac{K_1(a_i) - K_1(a_j)}{K_2(a_j) - K_2(a_i)} > 0. \tag{1.7}$$

Multiplicative metrized relations of linear order can be set on a set of criteria.

Decision-making is carried out on the basis of the choice of one of the alternative options e_i, $e_i \in E$. So, the choice of the optimal option is carried out using the criterion.

$$E_0 = \{E_{i0},\ E_{i0} \in E \wedge e_{i0} = \max_i e_i\} \tag{1.8}$$

The practical significance of this criterion for the problem under consideration is that the set E_0 contains a set of variants of E_{i0}, belonging to the set E, and the e_{i0} score of which is the maximum among all the e_i scores.

If there are multiple options for programs and training systems for the air transport industry, a particular E_i solution option may be affected by different external F_j conditions, and, thus, alternative results of e_{ij} solutions are obtained.

The resulting evaluation of the e_{ij} solution, which corresponds to the E_i variant and the F_j conditions, can emphasize the advantage of the training system in terms of utility, reliability and cost-effectiveness. [61–64].

To find the optimal conditions F_j that satisfy the variants of the solution E_i, it is necessary to enter an objective function in which the matrix of solutions e_{ij} will be reduced to a single column. This will make it possible to concretize the variants of E_i with some result of e_{ij}.

In this regard, the kind of evaluation function

$$\max_i e_{ir} = \max_i(\min_j e_{ij} + \max_j e_{ij}). \tag{1.9}$$

Thus, forming the desired result, the compromise between the optimistic and pessimistic approaches is taken into account.

Optimistic position

$$\max_i e_{ir} = \max_i(\max_j e_{ij}). \tag{1.10}$$

To build an optimistic position, the decision results matrix e_{ij} is used in which you should select the row with the maximum values from all possible results.

To build a neutral position of building a training system, can be used a model:

$$\max_i e_{ij} = \max_i(1/n \sum_{j=1}^{n} e_{ij}). \tag{1.11}$$

This model will allow you to estimate the minimum damage, taking into account the set of worst-case results.

Pessimistic position

$$\max_i e_{ij} = \max_i(\min_j e_{ij}). \tag{1.12}$$

The model allows you to take into account the unfavorable options for organizing the training system for each of the alternative options. Next, the most favorable option is selected; i.e., the best option is selected from the entire set of unfavorable ones.

Using criterion (1.8) and taking into account the combination of its largest and smallest result, the consequences of each of its alternative solutions can be represented

$$e_{ir} = \min_j e_{ij} + \max_j e_{ij}. \quad (1.13)$$

The formation of a set of alternative solutions follows directly from the problem statement.

In addition to the previously mentioned approaches, the optimization method and integer linear programming method can also be used in the formation of optimal solutions of the problem.

The task is formulated as follows.

$$\max(\min) : F(x) = \sum_{j=1}^{n} c_j x_j \quad (1.14)$$

under the conditions

$$\sum_{j=1}^{n} a_{ij} x_j \leq b_i, \ i = \overline{1, m}, \quad (1.15)$$

$$x_j \geq 0, \ j = \overline{1, n}, \quad (1.16)$$

$$x_j - integers, \ j = \overline{1, n_1}; \ n_1 \leq n. \quad (1.17)$$

If $n_1 = n$, then task (1.14)–(1.17) is fully integer; if $n_1 < n$, then task is a partially integer linear programming problem.

The set of vectors $X = (x_1, \ldots, x_n)$, satisfying the conditions (1.15)–(1.17), is the domain of the definition of the integer programming problem (1.14)–(1.17).

The vector X, satisfying the conditions (1.15)–(1.17), is a plan (or a valid solution) of the integer programming problem (1.14)–(1.17).

A plan $X^* = (x_1^*, \ldots, x_n^*)$, converting the linear form (1.14) to the maximum is the optimal plan (or solution) of the integer programming problem (1.14)–(1.17).

If $X = (x_1, \ldots, x_n)$ is a plan (optimal plan) and $x_0 = \sum_{j=1}^{n} c_j x_j$, then $\tilde{X} = (x_0, \ldots, x_n)$ is extended (extended optimal) plan of the integer programming problem (1.14)–(1.17).

The following symbols are introduced:

D is scope of the task definition (1.15)–(1.16);

D^μ is scope of the task definition (1.15)–(1.17);

(D, F) is the task (1.14)–(1.16);

(D^μ, F) is the task (1.14)–(1.17);

$X(D; F)$ is optimal plan of task (1.14)–(1.16);

$X(D^\mu, F)$ is optimal plan of task (1.14)–(1.17);

$\tilde{X}\ (D^\mu, F)$ is extended optimal plan of task (1.14)–(1.17).

An integer programming problem is solvable, if there is an optimal plan X^*.

Another area of application of integer models is the choice of options.

In the corresponding tasks, all or some variables can take only two values: 0 or 1; i.e., Boolean variables are used.

The mathematical formulation of the problem has the form.

$$F(x) = \sum_{j=1}^{n} c_j x_j \to \max, \quad (1.18)$$

$$\sum_{j=1}^{n} a_{ij} x_j \le b_i,\ i = \overline{1, m}, \quad (1.19)$$

$$x_j \in \{0, 1\}\ for\ j = \overline{1, n}.$$

Thus, the use of the methods of forecasting and expert optimization modeling, described above, will allow us to solve the problems of improving the efficiency of managing the training of specialists for work at civil aviation enterprises.

References

1. Vafin RK, Mosyagin GM, Nikonorov AV, Petin BF (1988) Scientific-methodical bases of formation of the purposes of training in a technical college. In: Questions to improve the educational process: proceedings of the Moscow higher technical school named after N. E. Bauman. Moscow: Moscow higher technical school named after N. E. Bauman, no 50, p 5–14
2. Dorrer AG, Ivanilova TN (2007) Modeling of interactive adaptive learning rate. Modern problems of science and education (5):52–59
3. Eliseev BP (2016) We want to make the University an «expert institute» of Federal air transport Agency. The system administrator, no 5, p 64
4. Krasovskii AA (1992) Aviation training devices. M: Air force engineering academy named after N. E. Zhukovsky, p 356
5. Latyshev VL (1982) Analysis of the structure of the course for automated learning systems. Automated systems for research, training and management in higher education: collection of scientific works. Novosibirsk: Novosibirsk State University, pp 26–32
6. Panteleev ER (1999) Metamodel of resource allocation in the task of designing the educational schedule of the university. Inf Technol (7):45–49
7. Krivov VD, Filatov YI (eds) (2007) The main directions of improving the training of flight and engineering personnel for civil aviation of the Russian Federation: collection of materials. Moscow, p 110
8. Fedotov AV (1985) Modeling in university management. L: Leningrad state university, p 120
9. Aidarkin DV (2012) Ensuring structural connectivity of modules of professional training of flight personnel for the purpose of increasing the level of flight safety: abstract of the PhD thesis. Saint-Petersburg, p 23
10. Devisilov VA (2008) On improving the training of personnel for safety in air transport. Scientific bulletin of the Moscow state technical university of civil aviation. no 135, p 7–21
11. Evdokimov VG (2013) Monitoring and ensuring flight safety, taking into account, changes in functional properties and risk factors of complex technical systems (aviation systems): autoreftrat of the dissertation of doctor of technical science. Moscow, p 38
12. Morozova OO (2014) Method for optimizing the process of managing aviation event risk factors, based on the minimum total cost criterion in the airline flight safety management system: autoreferat of the dissertation of the candidate of technical sciences. Ulyanovsk, p 22

13. Zairova ST (2014) Improving the professional psychophysiological selection of aviation specialists of the Republic of Kazakhstan: autoreferat of the dissertation of the candidate of technical sciences. Saint-Petersburg, p 21
14. Markova AK (1996) Psychology of professionalism. M: Znanie, p 308
15. Zakharova OG (2014) Improving the effectiveness of professional training of flight personnel on the basis of improving models and means of organizing training: autoreferat of the dissertation of the candidate of technical sciences. Ulyanovsk, p 22
16. Markova IA (2013) Modeling of the structure of the training course of individual trajectories of its passage. Vector of science of Tolyatti state university, no 3, p 443–445
17. Mikhalchevsky YY (2014) The method of professional training of pilots to make decisions in flight in order to improve the safety of aircraft flights: autoreferat of the dissertation of the candidate of technical sciences. Saint-Petersburg, p 24
18. Ovchinnikov AA (1982) Network methods of planning and organizing the educational process. M: Higher school of economics, p 160
19. Ostapchenko YB (2017) Models and means of training personnel of ground services of civil aviation to make decisions on the way out of emergency situations with the use of a complex automated system: autoreferat of the dissertation of the candidate of technical sciences. Saint-Petersburg, p 22
20. Krasnov SI (2013) The main directions of improving the training of aviation personnel in the Ulyanovsk higher aviation school of civil aviation (Institute) and its branches. In: Formation and development of branch science and education in the Russian air transport: collection of abstracts of scientific conference reports. Moscow, Federal state unitary enterprise state research institute of civil aviation, p 23–28
21. Kuznetsov YV (2013) Transformation of the domestic experience in the organization of professional training of aviation specialists (historical and pedagogical aspect): autoreferat of the dissertation of the candidate of technical sciences. Ulyanovsk, p 30
22. Smurov MY (2013) On the main directions of the development of professional education in civil aviation in the conditions of innovative formation and modernization of the Russian economy. In: Formation and development of branch science and education in Russian air transport: collection of abstracts of scientific conference reports. Moscow, Federal state unitary enterprise state research institute of civil aviation, p 19–23
23. Zhdanova MM, Preobrazhensky AP (2011) Questions of formation of professionally important qualities of an engineer. Bulletin of the Tajik technical university 4(4):122–124
24. Ivanov Yu S, Kiseleva NA, Sheibut YE (2000) Features of integration of educational disciplines in computerization of training. Specialist (3):27–28
25. Raven J (2002) Competence in modern society: identification, development and implementation. J Raven-M, Kogito-Center 400
26. Frolov Yu V, Makhotin DA (2004) Competence model as a basis for assessing the quality of training of specialists. J «Higher Educ Today» (108):34–41
27. Shmelkov AV (2006) Qualimetric monitoring of the results of educational activity within the competence model. Sci Bull MGTUGA (108):122–125
28. Borzova AS (2017) Calculation of the optimal labor intensity of the variable structure of the training course. Modeling and optimization in automated systems: materials of the All-Russian scientific school. Voronezh, Part 1, p. 31–37
29. Professional standard «Specialist in after-sales service of aviation equipment» (Order of the Ministry of Labor of the Russian Federation of 21.12.2015 no 1052n)
30. Gorlenko OA, Miroshnikov VV, Mozhaeva TP (2011) Coordination of bachelor's and master's competencies with the requirements of professional standards. In: Gorlenko OA (ed). Bryansk: Bryansk state technical university, p 211
31. Federal state educational standard of higher education in the field of training 250301. Technical operation of aircraft and engines (Order of the Ministry of education and science no. 1416 of 03.12.2015)
32. Federal state educational standard of higher education in the field of training 250302. Technical operation of aviation electrical systems and flight navigation systems (order of the Ministry of education and science no. 999 of 11.08.2016)

33. Federal state educational standard of higher education in the field of training 250401. Technical operation of aircraft and engines (Order of the Ministry of education and science no. 38854 of 09.09.2015)
34. Federal state educational standard of higher education in the field of training 250402. Technical operation of aviation electrical systems and flight navigation systems (Order of the Ministry of education and science no. 36957 of 21.04.2015).
35. Federal state educational standard of higher education in the field of training 250503. Technical operation of transport radio equipment (Order of the Ministry of education and science no. 43860 of 28.09.2016).
36. Federal state educational standard of higher education in the field of training 250504. Flight operation and application of aviation complexes (Order of the Ministry of education and science No. 43580 of 06.09.2016).
37. Federal state educational standard of secondary vocational education in the specialty 25.02.01. Technical operation of aircraft and engines (Order of the Ministry of education and science no. 389 of 22.04.2014)
38. Federal state educational standard of secondary vocational education in the specialty 25.02.02. Maintenance of aircraft with fuel and lubricants (Order of the Ministry of education and science no. 391 of 22.04.2014).
39. Federal state educational standard of secondary vocational education in the specialty 25.02.03 Technical operation of electrified and flight-navigation systems (Order of the Ministry of education and science no. 392 of 22.04.2014)
40. Federal state educational standard of secondary vocational education in the specialty 25.02.04 Flight operation of aircraft (Order of the Ministry of education and science no. 393 of 22.04.2014)
41. Federal state educational standard of secondary vocational education in the specialty 25.02.05 Air transport traffic management (Order of the Ministry of education and science no. 394 of 22.04.2014)
42. Brakhman TD (1984) Multicriteriality and the choice of alternatives in technology. M: Radio Commun p 288
43. Buslenko NP (1978) Modeling of complex systems. M: Nauka, p 400
44. Eliseev BP, Borzova AS, Koryakin ND (2016) State, prospects and problems of implementation of business process management of air enterprises. Innov Civil Aviation (4):5–16
45. Emelyanov SV, Larichev OP (1985) Multicriteria methods of decision-making. M.: Znanie, p 32
46. Leontiev LP, Gokhman OG (1984) Problems of educational process management: mathematical models. Riga: Zinatne, p 239
47. Lotov AV (2008) Multi-criteria problems of decision-making. M: MAKS Press, p 197
48. Chernorutsky IG (2004) Optimization methods in control theory: a textbook. St. Petersburg: Piter, p 265
49. Shannon K (1963) Works on information theory and cybernetics. M: Foreign literature, p 830
50. Glinsky VV, Ionin VG (1998) Statistical analysis. Training manual. Edition 2-e, revised and supplemented. M: Information and publishing house «Filin», p 264
51. Choporov ON, Zolotukhin OV, Manakin II, Bolgov SV (2015) The methodology of forming an information database for conducting multi-level monitoring and classification and prognostic modeling. Bulletin of the Voronezh institute of high technologies, no 14, pp 19–24
52. Novoseltsev VI (2003) System analysis: modern concepts. Voronezh: Publishing house «Kvarta», p 360
53. Tyurin YN, Makarov AA (2008) Analysis of data on a computer: a textbook. M: Publishing house «FORUM», p 368
54. Uemov VI (1978) System approach and general theory of systems. M: Naukap, p 272
55. Box J, Jenkins G (1974) Time series analysis. In: Pisarenko VF (ed) Forecast and management. M: Mir, no 1, p 406
56. Beshelev SD, Gurvich FG (1980) Mathematical and statistical methods of expert evaluation. M: Statistics, p 263

57. Beshelev SD, Gurvich FG (1973) Expert assessments. M: Nauka, p 59
58. Zadeh L (1976) The concept of a linguistic variable and its application to the adoption of approximate decisions—M. Mir, p 165
59. Berezovsky BA (1989) Multicriteria optimization. Mathematical aspects. M: Nauka, p 128
60. Beresov AV (1987) Multicriteria optimization in examples—M, Nauka, p 128
61. Gladun VP (1987) Planning solutions. Naukova dumka, Kiev, p 167
62. Andreatta G, Brunetta L, Odoni AR, Righi I, Stamatopoulos MA, Zografos KG (1999) A set of approximate and compatible models for airport strategic planning on airside and on landside. Air Traffic Control Q 291–317
63. Andersson T, Varbrand P (2004) The flight perturbation problem. Transp Plan Technol 91–117
64. Andreatta G (1987) Aircraft flow management under congestion. Transp Sci 249–253

Chapter 2
Conceptual Foundations of the Model-Based Approach to Forecasting and Optimization of the Training System in the Field of Air Transport Operation

A systematic analysis of the problems of managing the training of personnel with professional and higher education in the field of air transport maintenance shows that the basic mechanisms for improving the level of the personnel support system are [1, 2]:

ensuring that the volume characteristics of the personnel training system correspond to the needs of air enterprises;

accounting in educational resources of innovative and optimization approaches, focused on advanced requirements for the efficiency, quality and reliability of the functioning and development of air transportation;

achieving maximum convergence of the competence characteristics of professional and educational standards;

modernization of the industrial educational cluster, based on innovative personnel requirements and new generation standards.

The invariant of the implementation of these mechanisms is proposed to consider a number of models, on the one hand, allowing for a reliable forecast of the need for personnel, and on the other—determining the general basis for the characterization of labor functions, competencies and educational resources. This message is the main essence of the methodology of the study of the system of personnel support formation, using a model-oriented approach. Such an approach in this area with a focus on the management of personnel training has not been used and requires, first of all, a holistic conceptual framework in the form of a set of principles.

These principles are related to the variety of forecasting and optimization tasks, the multi-method nature of their construction and application, the need to combine formalized and expert methods of information processing and the multi-alternative nature of the decisions made. The integration of these principles is aimed at the optimal formation of the system of personnel training in the field of air transport maintenance, both in terms of volume indicators and in terms of model-oriented educational resources.

A. Borzova et al., *Modeling and Forecasting of Staffing in Civil Aviation*,
Springer Aerospace Technology, https://doi.org/10.1007/978-981-19-6238-7_2

2.1 The Principle of Duality and Multidimensionality of the Use of Models

This principle defines two directions in the use of models.

The first of them is related to the use of mathematical models for solving forecasting and optimization problems. At the same time, we divide the models into three groups:

predictive;
estimated values;
optimization tests.

In relation to the problem of improving the system of personnel training, predictive models allow us to determine the values of the volume characteristics of the need for personnel of air companies and training of specialists in universities on the basis of statistical information. Evaluation models are based on expert information and set priorities for the components of the modernization of the training system. The optimization models integrate formalized information that allows us to link the optimized variables with extreme and boundary requirements for the performance indicators of the training system [3].

At the same time, optimization modeling is a partial implementation of the duality principle in several aspects. The first aspect is determined by two mechanisms for achieving the effectiveness of the training system: reduction and transformation.

Under the reduction mechanism, we will understand the process of dividing the studied sets that characterize the training system into two subsets, one of which is more consistent with the extreme and boundary requirements for performance indicators.

Another form of reduction is the division of these sets within the framework of dual training [4], which provides a balance of supply and demand in labor market of the civil aviation. At the same time, it is necessary to divide the entire set of components into theoretical ones, which are taught in an educational institution, and practice-oriented ones, which are trained at existing air companies.

In the case where the components of the sets are models of professional activity, it is necessary to perform several reduction cycles. This mechanism will be called multi-stage reduction. Each stage is an evaluation simulation, based on group expert information. This procedure allows you to evaluate a promising set of components and exclude from further consideration the components that are not included in the number of promising ones. Evaluation modeling consists in ranking the elements of the original set by experts and setting the boundary value of the rank for the reduction.

The transformation mechanism is aimed at forming such a set of elements of the personnel training system that best shows the properties of other sets, defined by regulatory documents.

The general structure of optimization models that show the mechanisms under consideration is related to another aspect of the implementation of the duality principle by dividing the performance indicators of the training system into two groups that meet extreme and boundary requirements.

Full set of indicators y_g, $g = \overline{1, G}$.

To the part of these indicators $g_1 = \overline{1, G_1}$ extreme requirements are required:

$$y_{g_1} \to \text{extr}, g_1 = \overline{1, G_1}, \tag{2.1}$$

to the rest the boundary requirements are required::

$$y_{g_2} \le y_{g_2}^{\text{гр}}, g_2 = \overline{1, G_2}, \overline{1, G_1} \cup \overline{1, G_2} = \overline{1, G}. \tag{2.2}$$

These indicators depend on the choice of certain characteristics of the implementation of reduction and transformation mechanisms, which are formalized as a vector of optimized variables, measured on continuous and discrete scales

$$x = \left(x_1, \ldots, x_j, \ldots, x_J\right),$$

where $j = \overline{1, J}$—a numbered set of optimized variables.

This is another aspect of the duality of the two forms of realization of the vector x :

$x^1 = \left(x_1^1, \ldots, x_j^1, \ldots, x_{g_1}^1\right)$—vector of continuous variables;

$x^2 = \left(x_1^1, \ldots, x_j^2, \ldots, x_{j_2}^2\right)$—vector of discrete variables, in particular Boolean variables

$$x_j^2 = \begin{cases} 1, j = \overline{1, J_2}, \\ 0 \end{cases}.$$

In this case, the optimization model has the following form:

$$\begin{gathered} y_{g_1} = \Psi_{g_1}(x^1, x^2) \to extr, g_1 = \overline{1, G_1}, \\ y_{g_2} = f_{g_2}(x^1, x^2) \le y_{g_2}^{\text{гр}}, g_2 = \overline{1, G_2}, \\ x_j^{1min} \le x_j^1 \le x_j^{1max}, j = \overline{1, J_1}, \\ x_j^2 = \begin{cases} 1, \\ 0 \end{cases}, j = \overline{1, J_2}. \end{gathered} \tag{2.3}$$

Linear models with Boolean variables are of particular importance for optimizing the training system

$$\sum_{j=1}^{J_2} c_j x_j^2 \to \text{extr}, \sum_{j=1}^{J_2} a_{g_2 j} x_j^2 \le b_{g_2}, g_2 = \overline{1, G_2}, x_j^2 = \begin{cases} 1 \\ 0 \end{cases}, \quad j = \overline{1, J_2}, \tag{2.4}$$

where c_j—coefficients of a linear objective function,

$a_{g_2 j}$—coefficients of a linear system of constraints.

The next aspect is related to two mechanisms for organizing the optimization modeling process: randomization and smoothing. Randomization consists in the transition from the vectors x^1, x^2 to their probabilistic analogs. So, instead of the coordinates of the vector x_j^1, $j = \overline{1, J_2}$, in the process of optimization search on the kth iteration $k = 1, 2, 3\ldots$, random realizations $\tilde{x}_j^k$, $j = \overline{1, J}, \ldots$ are considered, which meet the following condition for a given type of distribution:

$$m(\tilde{x}_j) = x_j^k,$$

where $m(\cdot)$ is the designation of the mathematical expectation.

In the case of a vector of discrete variables and, in particular, Boolean variables, random variables $\tilde{x}_j^2$, $j = \overline{1, J}$, having a distribution:

$$P(\tilde{x}_j^2 = 1) = p_{x_j^2}, P(\tilde{x}_j^2 = 0) = q_{x_j^2}, p_{x_j^2} + q_{x_j^2} = 1. \tag{2.5}$$

are introduced.

The introduction of iteration over random optimized variables $\tilde{x}_j^2$ leads to another aspect of duality—dichotomous reduction, performed in automatic mode. The reduction of the variety of sets that characterize the training system is performed, if the probability values of combining promising subsets are close to 1, and non-promising subsets are close to 0. In the conditions of combinatorial uncertainty of the choice of perspective sets, a uniform distribution

$$p_{x_j}^1 = 0.5, q_{x_j}^1 = 0.5.$$

is assumed at the first step of dichotomous reduction

The change in the distribution (2.5) in the optimization search process is achieved by including in the iterative procedure predictive estimates A, calculated on the basis of smoothed transformations of the optimization model in the form of a dependency $F\left(x_j^2\right)$:

$$p_{x_j^2}^{k+1} = p_{x_j^2}^k + A\left(F\left(x_j^{2k}\right)\right). \tag{2.6}$$

The directed search in accordance with the distributions (2.6) leads to the synchronous formation of the preference ratio of the variants of the designed training system, which are quantified by the probability vector [5].

$$P = (p_1, \ldots, p_l, \ldots, p_L),$$

whose coordinates meet the condition

$$\sum_{l=1}^{L} p_l = 1.$$

To determine the dominance of certain options by the value of the probability of preference p_l, we use a function with the following properties [6]:

- symmetric with respect to the coordinates of the vector p, which means that it is independent of its relative position p_l and allows you to arbitrarily organize the correspondence of the variants of the system under study and the numbering set $l = \overline{1, L}$;
- reaching the maximum, when the options with the same probability of preference are equal

$$p_l^1 = \frac{1}{L};$$

- reaching the minimum, when dominating by the probability of preference of one option

$$p_v^2 = 1, \, p_l^2 = 0, l \neq v.$$

As such a function, we use one of the main information characteristics–entropy [7].

$$H(p) = -\sum_{l=1}^{L} p_l \lg p_l \tag{2.7}$$

in the sense that it characterizes the processes of reduction and transformation during the iterative search in the transition from the vector p^1 and the entropy $\left(p^1\right) = \lg L$ at the initial stage to the vector p^2 and entropy $H\left(p^2\right) = 0$ at the stage of final determination of the most promising (dominant) variant of the training system. In the use of the value (2.7), the ability to determine the values of p_l, based on the values of $p_{x_j}^2$, $j = \overline{1, J_2}$, calculated by the algorithm (2.6), also plays an important role [5].

The second direction is focused on creating a bank of professional activity models, on the one hand, related to the characterization of labor functions in professional standards, and on the other—showing in a formalized form the components of educational resources that are aimed at selecting certain competencies of personnel in the field of air transport maintenance.

The models of the first and second directions can be implemented by different methods, which complement the principle of duality of the model-oriented approach with the characteristic of multimethodality. The multi-method approach to the application of models dictates the need for a comparative analysis of the results of solving problem-oriented problems.

2.2 The Principle of Expert-Statistical Formation of Predictive and Assessment Models

When justifying the first principle, it is indicated that statistical (retrospective) information is used to build predictive models. The main form of statistical information that determines changes in the system of personnel training is the time series of volume indicators of the labor market demand for specialists with higher education and graduation from universities:

$$y_g(t_1), y_g(t_2), \ldots, y_g(t_k), \quad g = \overline{1, G}, \tag{2.8}$$

where $g = \overline{1, G}$—numbering set of indicators;

t_k—the time points, at which the values of the indicators are recorded with an interval Δt;

$k = \overline{1, K}$—a numbered set of time periods.

The task of data processing (2.8) is to build a mathematical model.

$$y_g = f_g(t), \tag{2.9}$$

which allows you to determine with a certain accuracy, computational cost and quality of the value $y_g(t)$ at the time point $(k+1), \ldots, (k+K)$, which follow the moment t_k, where K—the forecast period. To build a predictive model (2.9), it is proposed to use a number of time series processing methods (2.8) that are typical for technical systems [8–11].

These methods include the most common in time series forecasting. As a predictive analytical expression $f_g(t)$, a polynomial of the following form is usually used:

$$y_g(t) = f_g(t) = \sum_{h=1}^{H} A_{gh}\varphi_h(t),$$

where $\varphi_h(t)$—basic expressions that form the basis of the predictive function;

A_{gh}—adaptation coefficients;

$h = \overline{1, H}$—the number of basic expressions that allow you to make predictions with a certain accuracy.

The variety of predictive modeling, based on additive convolution of basic time functions, is associated with the possibility of using as $\varphi_h(t)$:

- polynomials of various degrees;
- orthogonal Chebyshev polynomials;
- empirical expressions that take into account most fully the specifics of changes in volume indicators in the training system.

Another method of predicting changes in indicators is the use of elementary functions:

– linear function

$$f_g(t) = a_{0g} + a_{1g}t;$$

– quadratic function

$$f_g(t) = a_{0g} + a_{1g}t + a_{2g}t^2;$$

– cubic function

$$f_g(t) = a_{0g} + a_{1g}t + a_{2g}t^2 + a_{3g}t^3;$$

– fractional–rational function, in particular, fractional–linear function

$$f_g(t) = \frac{a_{0g} + a_{1g}t}{b_{0g} + b_{1g}t};$$

– exponential function

$$f_g(t) = be^{at}.$$

To process information about the time series (2.8) in order to calculate the coefficients $A_{gh}, a_{0g}, a_{1g}, a_{2g}, a_{3g}, b_{0g}, b_{1g}, b, a$ the least squares and exponential smoothing methods are used [12].

Thus, the multi-method approach leads to the possibility of calculating predictive estimates with different quality indicators, which we will refer to as the most significant ones.

1. Prediction accuracy ε

$$\varepsilon_g = y_g^{np} - y_g^0,$$

where y_g^{np}—predictive assessment of gth indicator of the training system;
y_g^0—true value of gth indicator.

2. Performance of forecasting T^{fc}, characterizing the time spent on the forecasting process, or the ratio of the forecasting time T^{fc} to time, which is covered by the forecast $K\Delta t$, where Δt is the time interval between the forecast periods.
3. The cost of forecasting Z^{fc}, measured by the cost of computing resources for the implementation of the forecasting process.
4. Information indicator of the quality of forecasting, which indicates, how much information about the functioning of the training system has increased as a result of forecasting

$$\delta H = \frac{\sum_{g=1}^{G}\left(H_{0g} - H_g^K\right)}{\sum_{g=1}^{G} H_{0g}},$$

where H_{0g}—the value of the initial entropy by gth parameter,

$$H_{0g} = -\sum_{g=1}^{G} p_g^0 \lg p_g^0,$$

p_g^0—probability that characterizes the significance of the gth indicator at the initial stage of forecasting

$$\sum_{g=1}^{G} p_g^0 = 1;$$

$$H_g^K = -\sum_{g=1}^{G} p_g^K \lg p_g^K,$$

H_g^K—the value of entropy at the final stage of forecasting for Kth time interval;

p_g^K—the probability that characterizes significance gth indicator for Kth time interval

$$\sum_{g=1}^{G} p_g^K = 1.$$

5. The indicator of the completeness of forecasting, which characterizes the ratio of the number of assessments of the quality training system, for which it is possible to generate data for the predicted time series on the basis of statistical information, to the total number of indicators that determine the functioning of the system under study.

After receiving the results of forecasting using each of these methods, you should choose the result that most reliably shows the trends of changes of the volume indicators in the training system for air transport maintenance companies.

In this case, an evaluation model is built on the basis of expert information and the method that has the highest priority for this model is selected. In the process of individual and group expert evaluation, the priorities of a set of elements (alternatives) are determined:

$$W_1, \ldots, W_l, \ldots, W_L, \tag{2.10}$$

where $l = \overline{1, L}$—numbering set of elements (alternatives), considered by the experts. There are a number of methods of conducting exams, which are divided into three groups: organization of exams, processing of exam results, selection of the highest priority element from the set (2.10) [13–18].

As a method of organizing expert examinations by a group of $d = \overline{1, D}$ experts, it is proposed to use an open discussion on the evaluation of indicators for ranking. The method of processing the results of the obtained expert assessments can be the ranking of indicators by each of the experts and bringing them further to the average value [19]. As a result, we have integer values of the ranks $r_g, g = \overline{1, G}$, which characterize the priorities of the indicators $y_g, g = \overline{1, G}$.

To select a priority option from the set (2.10), consider an adaptive human–machine procedure, based on the randomization and smoothing mechanisms mentioned above.

Randomization is determined by the introduction of the following discrete random variables and their distributions:

- instead of the numbering set $l = \overline{1, L}$, a discrete random variable $\tilde{l}$ is introduced, taking the values $\overline{1, L}$ with probability $p_l, l = \overline{1, L}, \sum_{l=1}^{L} p_l = 1$;
- instead of the numbering set $g = \overline{1, G}$, a discrete random variable $\tilde{g}$ is introduced, taking the values $\overline{1, G}$ with the probability $p_g, g = \overline{1, G}, \sum_{g=1}^{G} p_g = 1$;
- instead of the numbering set $d = \overline{1, D}$, a discrete random variable $\tilde{d}$ is introduced, taking the values $\overline{1, D}$ with the probability $p_d, d = \overline{1, D}, \sum_{d=1}^{D} p_d = 1$.

Building an adaptive procedure in a randomized environment requires to set the initial distributions of the listed random variables. As the initial distribution of the random variable $\tilde{l}$, in the absence of a priori information, before the experts evaluate the results of the prediction according to indicators (2.1)–(2.2), we assume a uniform distribution:

$$p_l^1 = \frac{1}{L}, \quad l = \overline{1, L}.$$

To calculate the initial distribution of the random variable $\widetilde{g}$, we use the values of the ranks of the indicators (2.1)–(2.2) $r_g, g = \overline{1, G}$.

$$p_g^1 = \frac{\frac{1}{r_g}}{\sum_{g=1}^{G} \frac{1}{r_g}}, \quad g = \overline{1, G}.$$

The initial distribution of the random variable $\tilde{d}$ is associated not only with the average values of the ranks r_g, $g = \overline{1, G}$, but also with the ranks r_g^d, assigned to the gth indicator by the dth expert:

$$p_d^1 = \frac{\sum_{g=1}^{G} \frac{1}{\left(r_g - r_g^d\right)^2}}{\sum_{d=1}^{D} \sum_{g=1}^{G} \frac{1}{\left(r_g - r_g^d\right)^2}}, \quad d = \overline{1, D}.$$

As a result of the adaptive process, using the values of the probabilities $p_l^k, l = \overline{1, L}$; $p_g^k = \overline{1, G}$; $p_d^k = \overline{1, D}$ at the kth step, it is necessary to determine the new values of these probabilities at the $(k+1)$th step, using the information, received from experts during the dialogue. The effectiveness of such an adaptive process is determined by the smoothed function, that is, the mathematical expectation of the set of values of indicators for the distribution of a random variable $\tilde{g}$ at the kth step

$$F_l = \sum_{g=1}^{G} p_g^k \hat{y}_g(W_l), \quad l = \overline{1, L},$$

where $\hat{y}_g(W_l)$ is the value of the gth indicator for the alternative W_l, normalized on the interval [0, 1],

$$\hat{y}_g(W_l) = \frac{y_g(W_l) - \min\limits_{l=\overline{1,L}} y_g(W_l)}{\max\limits_{1=\overline{1,L}} y_g(W_l) - \min\limits_{l=\overline{1,L}} y_g(W_l)},$$

$\min\limits_{l=\overline{1,L}} y_g(W_l)$, $\max\limits_{l=\overline{1,L}} y_g(W_l)$—the minimum and maximum values of the gth exponent on the numbering set of alternatives $l = \overline{1, L}$, respectively,
$y_g(W_l)$ is the value of gth exponent for alternative W_l.

1. By comparing the probabilities p_g^k with the value $\tilde{\xi}$, uniformly distributed over the interval [0, 1], we determine the value of a discrete random variable $\tilde{d}$ [20].
   If $p_1^k \le \tilde{\xi}$, then $d = 1$;
   if $p_1^k > \tilde{\xi}$, then $p_1^k + p_2^k$ is determined;
   if $p_1^k + p_2^k \le \tilde{\xi}$, then $d = 2$;
   if $p_1^k + p_2^k > \tilde{\xi}$, then $p_1^k + p_2^k + p_3^k$ is determined.
   The comparison is repeated until the implementation $\tilde{d} = \tilde{d}^k$ is determined. In the dialog an expert with the number d^k participates at the kth step.
2. By analogous comparison of probabilities p_l^k with a value $\tilde{\xi}$, the value of a discrete random variable $\tilde{l} = l^k$ is determined.

Next, an alternative W_{l^k} is presented to the expert d^k for evaluation.

3. The following survey is organized: "Which of the indicators $y_g(W_{l^k})$ does not satisfy the expert to the greatest extent?" We get the answer: "The exponent g^k". The survey continues: "To what extent is it necessary to change the value of the indicators $y_g(W_{l^k}),\ g = \overline{1, G}$ in the gradations of the linguistic variable 'it is necessary to change' in the direction of their desired improvement?" Gradations are used: strongly, significantly, somewhat, slightly, little [21–25].
4. Quantitative estimates of the numbering set $g_1^k, \ldots, g_s^k$ are formed, which includes the numbers of indicators from the numbering set $g = \overline{1, G}$, for which the linguistic variable takes the value <strong> :

$$\theta_g^k = \begin{cases} 1, & \text{if} g \in g_1^k, \ldots, g_s^k, \\ 0, & \text{otherwise}, \ g = \overline{1, G}; \end{cases}$$

$$\mu^k = \frac{\sum_{g=g_1^k}^{g_s^k} \mu_g^k}{S},$$

where μ_g^k is the value of the membership function, calculated for gth exponent by the values of linguistic variables [26];

μ^k—the average value of the membership function on the numbering set $g_1^k, \ldots, g_s^k$.

5. The number of indicators T_{gl}^k, for which the linguistic variable takes the value <small> in the case of an alternative $W_l,\ \ l = \overline{1, L}$ is determined.
6. The information, obtained in the previous stages of the dialogue with the expert, allows you to determine the distribution of a random variable $\widetilde{g}$for$(k+1)$th iteration

$$p_g^{k+1} = \frac{p_g^k \frac{1}{S} \chi\left(\theta_g^k\right) \varepsilon^{k+1}}{1 + \varepsilon^{k+1}}, \quad g = \overline{1, G},$$

where ε^{k+1}—value of step for calculating the distribution of a random variable $\widetilde{g}$ for $(k+1)$th iteration

$$\varepsilon^{k+1} = \varepsilon^k \exp\left[\frac{1}{S} \sum_{g=g_1^k}^{g_s^k} \operatorname{sign}\left(\theta_g^{k-1} \cdot \theta_g^k\right)\right],$$

$$\chi(a) = \begin{cases} 1, & \text{if}\, a > 0, \\ 0, & \text{if}\, a < 0. \end{cases}$$

7. The significance of alternatives $W_l,\ \ l = \overline{1, L}$ is estimated by a discrete value

$$\theta_l^k = \begin{cases} 1, \text{ if } \sum_{g=1}^{G} p_g^k \hat{y}_g^k(w_l) = \max_{l \in \overline{1,L}} \sum_{g=1}^{G} p_g^k \hat{y}_g^k(W_l), & l = \overline{1,L}. \\ -1 & \text{otherwise,} \end{cases}$$

8. The obtained information allows you to determine the distribution of a random variable $\tilde{l}$ for $(k+1)$th iteration

$$p_l^{k+1} = \frac{p_l^k \chi\left(\theta_l^k\right) \gamma_l^{k+1}}{1 + \gamma_l^{k+1}}, \quad l = \overline{1,L},$$

where γ_l^{k+1}—value of step for calculating the distribution of a random variable $\tilde{l}$for $(k+1)$th iteration

$$\gamma_l^{k+1} = \gamma_l^k \exp\left[\operatorname{sign}\left(\theta_l^{k-1} \cdot \theta_l^k\right)\right].$$

9. We keep the distribution of a random variable $\tilde{d}$ as constant during the dialog cycle

$$p_d^k = p_d^1, \; d = \overline{1,D}.$$

10. We select the best prediction option from the set (2.10):
    (a) the numbering set $l_1 = \overline{1,L_1} \in \overline{1,L}$ is determined, elements of the set (2.10), corresponding which, meet the condition $\max_l T_l^k$;
    (б) the numbering set $l_2 = \overline{1,L_2} \in \overline{1,L}$ is determined, elements of the set (2.10), corresponding which, meet the condition

$$\max_l \sum_{g=1}^{G} p_g^k y_g^k(W_l);$$

    (в) the numbering set as intersection the sets l_1 and l_2 is formed

$$l_3 = \overline{1,L_3} = l_1 \cap l_2 \in \overline{1,L};$$

    (г) an element from the set $l_3 = \overline{1,L_3}$ is selected as the best element W_l, according to expert estimates, and meet the condition

$$\max_{l_3} \sum_{g=1}^{G} p_g^k y_g^k(W_l).$$

An effective combination of modeling methods, based on statistical and expert information, ensures the implementation of the principle under consideration.

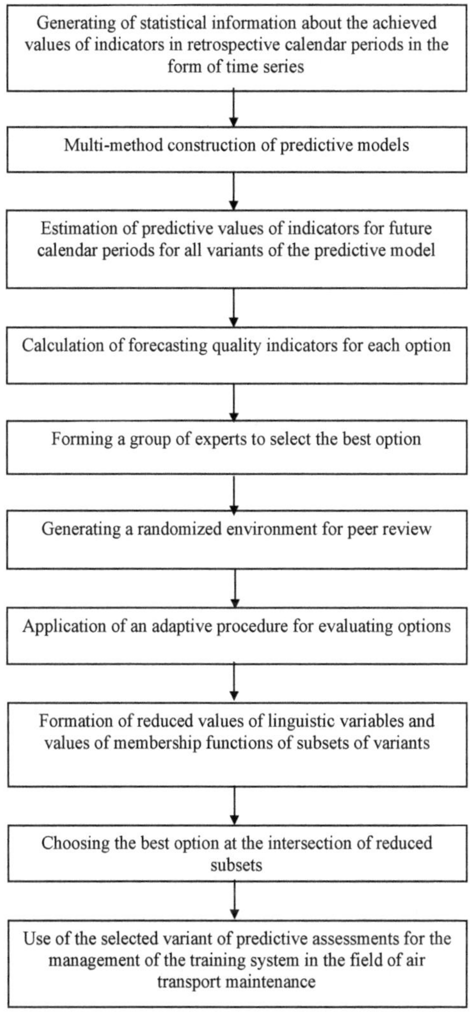

Fig. 2.1 Block diagram of expert-statistical formation of predictive and evaluation models

A block diagram of the expert-statistical formation of predictive and evaluation models is shown in Fig. 2.1.

2.3 The Principle of Multi-alternative Choice Models of Professional Activity

Let us consider professional activity through the following categories: types of activity, labor functions, professional competencies and corresponding models of professional activity.

The field of maintenance of air transport is characterized by:
types of activities

$$D_1, \dots, D_m, \dots, D_M, \tag{2.11}$$

where $m = \overline{1, M}$—numbering set of types of activities.
Each mth type of activity corresponds to:
labor functions in the professional standard,

$$\tau_{1m,}, \dots, \tau_{im}, \dots, \tau_{Im}, \tag{2.12}$$

where $i_m = \overline{1, I_m}$—numbering set of labor mth type of activity, based on knowledge and skills, for achievement which it is necessary to use a set of models

$$\mu_{1m}, \dots, \mu_{jm}, \dots, \mu_{Jm}, \tag{2.13}$$

where $j_m = \overline{1, J_m}$—numbering set of professional activity models;
competences in the educational standard

$$K_{1m}, \dots, K_{rm}, \dots, K_{Rm}, \tag{2.14}$$

where $r_m = \overline{1, R_m}$—the numbering set of competencies, required to perform mth type of activity and corresponding models

$$\mu_{1m}, \dots, \mu_{sm}, \dots, \mu_{Sm}, \tag{2.15}$$

where $s_m = \overline{1, S_m}$—numbering set of professional activity models.
The sets (2.11)–(2.15) will be formed using multi-stage reduction:
stage 1—selection from a set of (2.11) types of activities of promising elements that most determine the leading requirements for the efficiency of the air transport maintenance process;
stage 2—reduction of the full set of labor functions, included in the professional standard, according to the correspondence of a promising subset of directions;
stage 3—reduction of the set of professional activity models that correspond to the knowledge and skills that support the labor functions, identified at stage 2;
stage 4—reduction of the full set of competencies, included in the educational standard, according to the correspondence of a promising subset of areas;
stage 5—reduction of the set of professional activity models that correspond to the knowledge and skills that support the competencies, identified at stage 4;
stage 6—reduction based on the multi-alternative optimization model, which makes it possible to form a reduced set of professional activity models, identified at the 3rd and 5th stages;

stage 7—reduction of the subset, obtained at stage 6 by identifying invariant mathematical descriptions that characterize the entire reduced set of professional activity models.

As invariant mathematical descriptions, we will use the following:

- structural and logical models;
- situational models;
- optimization models.

The structural-logical model defines the composition of the elements that characterize the components of professional activity, and the connections between them in the form of hierarchical, tree-like and graph schemes, as well as logical relationships between the elements, using the apparatus of classical logic of statements and predicates and non-classical inductive logic.

The situational model describes the process states, that characterize the components of professional activity and the solutions that allow you to move from the initial state to the states with the best performance and quality indicators. The optimization model characterizes the relations between indicators and parameters in the form of mathematical statements (2.3), (2.4).

The multi-alternative choice corresponding to the 6th stage consists in determining such a set of models from the sets (2.13), (2.14), which in terms of the number of elements was minimal that corresponds to a limited educational resource, but at the same time each labor function (2.12) and competence (2.14) for all types of activity (2.11) corresponded to at least one element from this set.

We introduce variables that characterize the multi-alternative choice of models

$$x_{jm} = \begin{cases} 1, & \begin{array}{l}\text{if for formation of labor function } D_m\text{th} \\ \text{type of activity the model } \mu_{jm} \text{is used,}\end{array} \\ 0, & \text{in otherwise} \end{cases}, \; j_m = \overline{1, J_m},\, m = \overline{1, M}; \tag{2.16}$$

$$x_{sm} = \begin{cases} 1, & \begin{array}{l}\text{if for formation of competences of } D_m\text{th} \\ \text{type of activity the model } \mu_{sm} \text{ is used,}\end{array} \\ 0, & \text{in otherwise,} \end{cases}, \; s_m = \overline{1, S_m}, m = \overline{1, M}. \tag{2.17}$$

Then, the mathematical formulation of the optimization problem on the sets of variables (2.16), (2.17) for the D_mth type of activity has the form

$$\sum_{j_m=1}^{J_m} x_{jm} + \sum_{s_m=1}^{S_m} x_{sm} \rightarrow \min, \tag{2.18}$$

$$\sum_{j_m=1}^{J_m} c_{ijm} x_{jm} \geq 1, \quad i_m = \overline{1, I_m}, \tag{2.19}$$

$$\sum_{s_m=1}^{S_m} c_{rsm} x_{sm} \geq 1, \quad r_m = \overline{1, R_m}, \tag{2.20}$$

$$x_{jm} = \begin{cases} 1 \\ 0 \end{cases}, \; j_m = \overline{1, J_m}, x_{sm} = \begin{cases} 1 \\ 0 \end{cases}, \; s_m = \overline{1, S_m}, m = \overline{1, M} \tag{2.21}$$

In the optimization model (2.18)–(2.21), the optimization criterion determines the desire for a set of models with a minimum number of elements, provided that the constraints (2.19)–(2.21) are met. To formalize the constraints on the support of the models (2.13) for the labor functions $\tau_{im}, i_m = \overline{1, I_m}$, based on the expert assessment, the coefficients are introduced

$$c_{ijm} = \begin{cases} 1, & \text{if it is decided, based on the examination, that the model} \mu_{jm} \\ & \text{supports the formation of the labor functions} \tau_{im}, \\ 0, & \text{in otherwise,} \end{cases}$$
$$j_m = \overline{1, J_m}, i_m = \overline{1, I_m}, m = \overline{1, M},$$

and for restrictions (2.20) on the support of (2.15) competence models $K_{rm}, \quad r_m = \overline{1, R_m}$, coefficients are introduced

$$c_{rsm} = \begin{cases} 1, & \text{if it is decided, based on the examination, that the} \\ & \text{model} \mu_{sm} \text{ supports the formation of competences } K_{rm}, \\ 0, & \text{in otherwise,} \end{cases}$$
$$s_m = \overline{1, S_m}, r_m = \overline{1, R_m}, m = \overline{1, M}.$$

The model (2.18)–(2.21) is a classical linear model of discrete optimization—the task about minimum coverage, for which approximate methods of solving are developed [19, 26–29]. In the case of the algorithm (2.6), it is necessary to form the function $F(x_{jm}, x_{sm})$ by switching from the optimization problem (2.18)–(2.21) with constraints to the equivalent task without constraints [27].

$$F(x_{jm}, x_{sm}) = -\sum_{j_m}^{J_m} x_{jm} \sum_{s_m}^{S_m} x_{sm} + \sum_{i_m}^{I_m} \alpha_{im} \left(\sum_{s_m}^{S_m} c_{rsm} x_{sm} - 1 \right)$$
$$+ \sum_{r_m=1}^{R_m} \alpha_{rm} \left(\sum_{j_m=1}^{J_m} c_{ijm} x_{jm} - 1 \right), \tag{2.22}$$

where α_{im}, α_{rm}—coefficients, whose numerical values are $\alpha_{im} \geq 0, \alpha_{rm} \geq 0$ and determined at the specified initial values $\alpha^1_{im}, \alpha^1_{rm}$, based on algorithmic procedures

$$\alpha_{im}^{k+1} = \max\left\{0, \alpha_{im}^{k} - \beta_{im}^{k+1}\left(\sum_{jm=1}^{J_m} c_{ijm}x_{jm} - 1\right)\right\},$$

$$\alpha_{rm}^{k+1} = \max\{0, \alpha_{rm}^{k} - \beta_{rm}^{k+1}\left(\sum_{s_m}^{S_m} c_{rsm}x_{sm} - 1\right),$$

$\beta_{im}^{k} \geq 0, \beta_{rm}^{k} \geq 0$—the value of the steps, set by the expert advisor.

Using them, we obtain reduced sets of models for D_mth type of activity $\mu_{jm}^{*}, \mu_{sm}^{*}$ and form a single set for all types of activities $m = \overline{1, M}$

$$\mu = \mu_{jm}^{*} \bigcup_{m=1}^{M} \mu_{sm}^{*}.$$

We introduce a new numbering of the elements of the set

$$\mu_1, \ldots, \mu_n, \ldots, \mu_N \tag{2.23}$$

and let us move on to the further use of the model-oriented approach.

2.4 The Principle of Optimal Transformation of Professional Activity Models into Components of Educational Programs

The principle is based on an optimization model that links several models of professional activity (2.23) with several components of educational programs.

$$O_1, \ldots, O_v, \ldots, O_V, \tag{2.24}$$

where $v = \overline{1, V}$—the numbering set of components of educational programs.

The task of transformation is to effectively show the elements of the set (2.23), when forming the elements of the set (2.24). To give the problem an optimization character, we introduce alternative variables:

$$x_{nv} = \begin{cases} 1, & \text{if } \mu_n\text{th model is used for formation of} \\ & O_v\text{th component of the educational program,} \\ 0, & \text{inotherwise,} \end{cases} \quad n = \overline{1, N}, v = \overline{1, V}. \tag{2.25}$$

Next, using group expert evaluation, we will determine the rank sequence of models $\mu_n, n = \overline{1, N}$, based on the a priori ranking method [19, 30–33] to form the component of the educational program $O_v - a_{nv}$, where a_{nv} accepts integer values on the set $\overline{1, N}$ for the most significant model $a_{nv} = 1.a_{nv} = 1$.

Then, as an optimization criterion, we take the requirement to achieve the minimum of the function that depends on the variables (2.25),

$$\sum_{n=1}^{N}\sum_{v=1}^{V} a_{nv}x_{nv} \to \min, \tag{2.26}$$

that determines the use of the most significant models from the set (2.23) in the educational program. Limitations are associated with a limited time resource for the formation of labor functions and competencies using the component O_n:

$$\sum_{n=1}^{N} t_{nv}x_{nv} \leq \varepsilon_v T_v, \quad v = \overline{1, V}, \tag{2.27}$$

where t_{nv}—the labor intensity of the study μ_nth model as part of the educational program component O_v;

T_v—the labor intensity, determined by the educational plan for the learning the program component O_v;

ε_v—the coefficient, that characterizes the share of time, required to study models in the total labor intensity T_v.

Combining the optimization criterion (2.26) and constraints (2.25), (2.27) leads to the classical model of discrete optimization—a model of linear Boolean programming, which can be solved by approximate methods [19]:

$$\sum_{n=1}^{N}\sum_{v=1}^{V} a_{nv}x_{nv} \to \min, \tag{2.28}$$

$$\sum_{n=1}^{N} t_{nv}x_{nv} \leq \varepsilon_v T_v, v = \overline{1, V},$$

$$x_{nv} = \begin{cases} 1, \\ 0, \end{cases} n = \overline{1, N}, v = \overline{1, V}.$$

To use the algorithm (2.8), we form the function $F(x_{nv})$ by moving from the optimization problem (2.6) with constraints (2.25) and (2.27) to the equivalent problem without constraints

$$F(x_{nv}) = -\sum_{n=1}^{N}\sum_{v=1}^{V} a_{nv}x_{nv} + \sum_{v=1}^{V} \alpha_v \left(\varepsilon_v T_v - \sum_{n=1}^{N} t_{nv}x_{nv} \right),$$

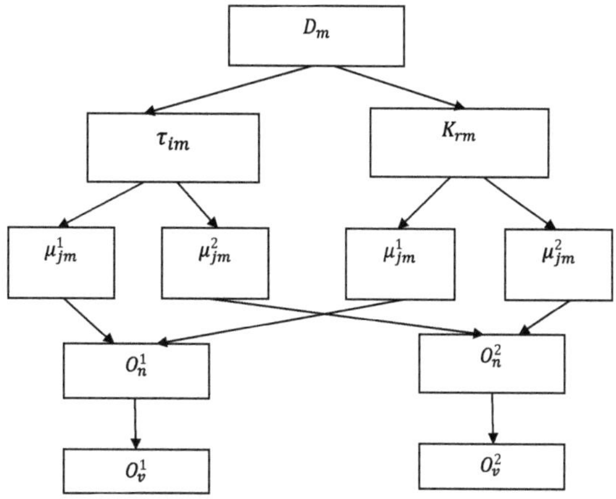

Fig. 2.2 Tree-like scheme of multi-stage reduction and transformation in the case of dual learning

where α_v—coefficients, whose numerical values are $\alpha_v \geq 0$ and determined at the specified initial values α_v^1, based on an algorithmic procedure

$$\alpha_v^{k+1} = \max\{0, \alpha_v^k - \beta_v^{k+1}\left(\varepsilon_v T_v - \sum_{n=1}^{N} t_{nv} x_{nv}\right),$$

where $\beta_v^k \geq 0$—the value of the steps, set by the expert.

As a result, a bank of educational resources is formed in the form of the most appropriate distribution of a set of professional activity models $\mu_n n = \overline{1, N}$ between components of the educational programs O_v, $v = \overline{1, V}$.

In the case of dual training, which requires the selection of theoretical and practice-oriented components, multi-stage reduction and transformation are carried out, using the tree scheme shown in Fig. 2.2. In this diagram, the models of professional activity are divided into two groups:

μ^1_{jm}, μ^1_{sm}—correspond to theoretical training;
μ^2_{jm}, μ^2_{sm}—correspond to practice-oriented training.

Accordingly, each of these groups of models is reduced to a single set μ_n^1, μ_n^2 using the optimization model (2.18)–(2.21), which in turn are transformed into many components of educational programs O_v^1, O_v^2, based on the optimization model (2.28).

Thus, the principles and procedures for their implementation, discussed above, form the basis of the model-oriented approach.

A block diagram of the use of the principles of the model-based approach for forecasting and optimizing the training system in the field of air transport maintenance is shown in Fig. 2.3.

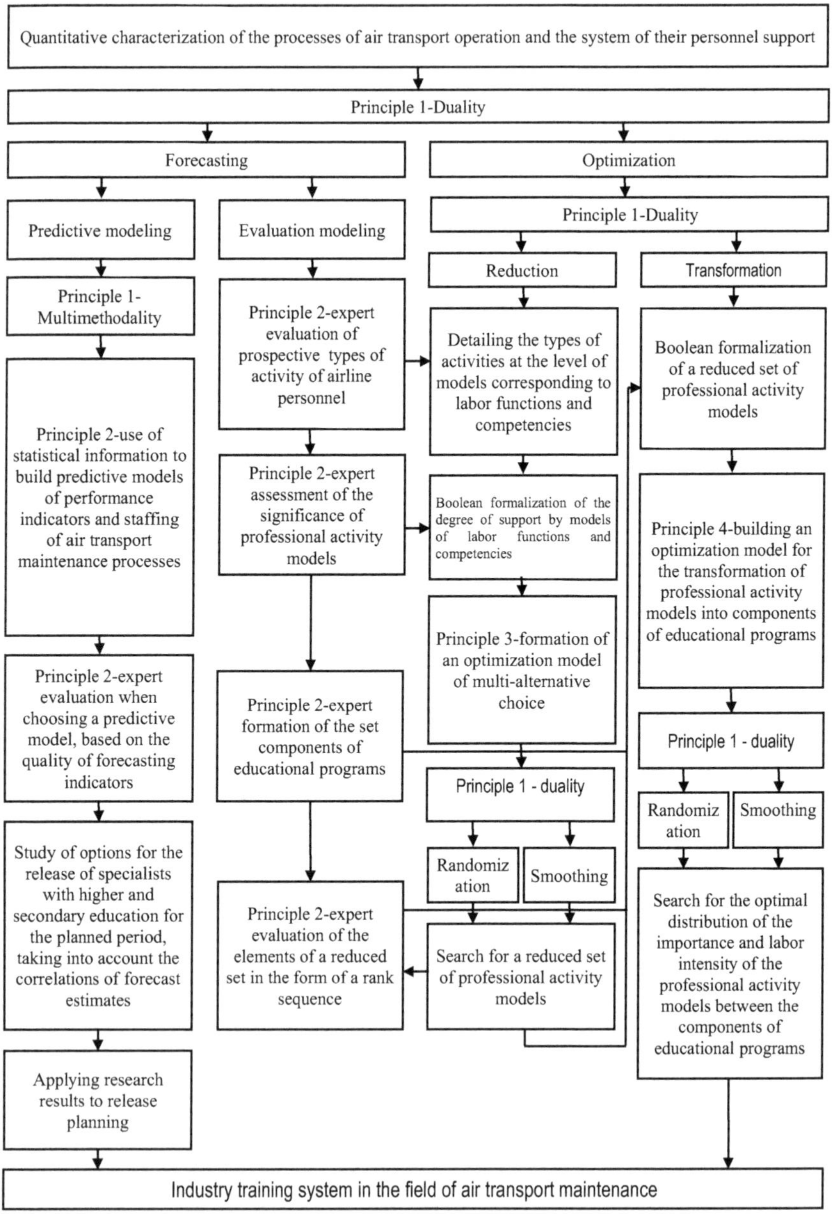

Fig. 2.3 Block diagram of the use of the principles of the model-based approach for forecasting and optimizing the training system in the field of air transport maintenance

References

1. Eliseev BP (2011) Methodology of system-technical management of professional education in civil aviation: abstract of dissertation of Doctor of technical science. Ufa, p 32
2. Celeberda IN, Neguritsa DS, Sokol VV (2008) The prospect of implementing an individual trajectory of training for storages with SCORM-courses. Educ Technol Soc 11(3):394–406
3. Eliseev BP (2010) Strategic tasks of the state in the field of higher education as optimization tasks. In: Theory and practice of social development, vol 4, pp 140–149
4. Lushnikov AS, Tarasov SN (2012) Ground-based radio-electronic means of ensuring aircraft flights: textbook. Ulyanovsk Institute of Civil Aviation, Ulyanovsk, p 46
5. Lvovich YaE (2006) Multi-alternative optimization: theory and applications. Kvarta, p 428
6. Eliseev BP, Liu D (2015) General principles and features of transport university management. In: Scientific bulletin of the Moscow state technical university of civil aviation, vol 216, pp 30–34
7. Afanasyev VG (2009) Fundamentals of managerial activity. Part 2. Practical aspects of aviation enterprise management. M.: MSTU GA, p 69
8. Borzova AS (2017) Development of airport infrastructure on the basis of optimization modeling of professional activity of an aviation radio engineer. Econ Manage Manage Syst 26(4):94–100
9. Andreatta G, Lorenzo B, Guastalla G (1998) The flow management problem: recent computational algorithms. Control Eng Pract 727–733
10. Barnhart C, Belobaba P, Odoni AR (2003) Applications of operations research in the air transport industry. Transp Sci 368–391
11. Bensaou M, Venkatraman N (1995) Configurations of inter organizational relationships: a comparison between US and Japanese automakers. Manage Sci 41:1471–1492
12. Box J, Jenkins G (1974) Time series analysis. In: Pisarenko VF (ed) Forecast and management (1). M: Mir, p 406
13. Litvak BG (1982) Expert information: methods of obtaining and analyzing. Radio and Communication, M., p 104
14. Litvak BG (1996) Expert assessments and decision-making. Patent, M., p 284
15. Lvovich YaE, Lvovich IYa (2010) Decision-making in the expert-virtual environment. Scientific Book, Voronezh, p 140
16. Berge M (1994) Timetable optimization: formulation, solution approaches, and computational issues. AGIFORS Proc 341–357
17. Blumstein A (1959) The landing capacity of a runway. Oper Res 752–763
18. Davidson WH (1993) Beyond re-engineering: the three phases of business transformation. IBM Syst J 32(1):65–80
19. Lvovich IYa, Lvovich YaE, Frolov VN (2016) Information technologies of modeling and optimization: brief theory and applications. Scientific Book, Voronezh, p 444
20. Sobol IM (1973) Numerical methods of Monte Carlo. M.: Nauka, p 312
21. Mansurov AN, Mansurova GM, Saveleva GP (2001) Application of the system-functional method for the analysis and evaluation of the professional activity of the university. In: Problems of the quality of education: collection of works XI all-Russian scientific and practical conference. Moscow, pp 27–29
22. Dear R, Sherif YS (1991) An algorithm for computer assisted sequencing and scheduling of terminal area operations. Transp Res 129–239
23. Dell'Olmo P, Lilli G (2003) A new hierarchical architecture for air traffic management optimisation of airway capacity in a free flight scenario. Eur J Oper Res 179–193
24. Desaulniers G, Solomon MM (1997) Daily aircraft routing and scheduling. Manage Sci 841–855
25. Dorndorf U, Drexl A, Nikulin Y, Pesch E (2007) Flight gate scheduling, state-of-the-art and recent developments. Omega 326–334
26. Beresov AV (1989) Multicriteria optimization in examples. M.: Nauka, p 128
27. Barnhart C, Boland N, Clarke L, Johnson E, Nemhauser G, Shenoi R (1998) Flight string models for aircraft fleeting and routing. Transp Sci 208–220

28. Gibson CF, Nolan R (1974) Managing the four stages of EDP growth. Harvard Bus Rev 52(1):76–89
29. Henderson JC, Venkatraman N (1993) Strategic alignment: leveraging information technology for transforming organizations. IBM Syst J 32(1):4–17
30. Leonardi PM (2007) Activating the informational capabilities of information technology for organizational change. Organ Sci 18(5):813–831
31. Liu R, Sun K (2010) Connotation and denotation of engineering management informatization. Sci Technol Progress Policy 27(19):1–4
32. Liu R, Sun K (2011) Research on architecture of engineering management informatization. Eng Sci 13(8):4–9
33. Mani D, Barua A, Whinston A (2010) An empirical analysis of the impact of information capabilities design on business process outsourcing performance. MIS Q 34(1):39–62

Chapter 3
The Formation of Predictive and Evaluative Models of Volume Training and Skills to Meet the Requirements Leading the Development of Air Transport

The development of the conceptual framework of the model-based approach to improving the system of personnel training for airlines allows us to move on to solving a number of problem-oriented tasks, related to the staffing support of the industry. The primary task is to assess the prospective values of the volume indicators of the output of specialists within the framework of the civil aviation educational cluster. When evaluating these indicators, it is not enough to focus only on the capabilities and effectiveness of educational organizations. The advanced requirements of air transport development determine the need for additional accounting of information, based on predictive models of airline performance indicators. An important role in this study is played by the use of GIS-oriented information about the geographically distributed airport system.

The formation of predictive models, based on spatiotemporal information resources, allows us to justify a formalized approach to geographically distributed planning of employment of graduates of industrial educational organizations in air transport enterprises. The basis for quantitative assessment and optimization in this case is a matrix of paired correlations between the indicators of the air transport system and the graduation volumes for the projected period of time.

The choice of a certain horizon for planning graduation and employment of graduates determines the need to determine the priority types of professional activities for this period, for which it is primarily necessary to have model-oriented educational resources. In this case, it becomes appropriate to use expert evaluation.

A. Borzova et al., *Modeling and Forecasting of Staffing in Civil Aviation*,
Springer Aerospace Technology, https://doi.org/10.1007/978-981-19-6238-7_3

3.1 The Predictive Modeling System Communications Personnel Software Industry and Performance Indicators the Development of Air Transport using a Spatiotemporal Data

The management decisions made in the system of training personnel for civil aviation depend not only on the correct assessment of the current situation, but also on the short-term forecast of changes in the analyzed indicators [1, 2].

It should be noted that a short-term forecast is a forecast "for tomorrow", i.e., for several steps ahead (usually up to 5 steps, in our case up to 5 years). To construct such a forecast, a time series model, based on the exponential smoothing method, is proposed for use [3–7]. The main advantage of this method is adaptability, which means that it is possible to take into account the information, contained in the time series levels in a differentiated way, in which the "older" information has less value in forecasting.

For a time series, an additive model of its representation is used:

$$y_t = m + \varepsilon_t, \tag{3.1}$$

where y_t—the value of the indicator under study at the moment of time t;

$m = \overline{y_t}$—mathematical expectation (average value);

ε_t—some random component.

The recurrent formula is used to smooth the series levels y_t

$$S_t = \alpha\, y_t + (1 - \alpha)\, S_{t-1}, \quad t = \overline{1,N}, \tag{3.2}$$

where α—smoothing coefficient ($0 < \alpha < 1$);

S_t—the smoothed value of the real level y_t at the moment of time t;

N—number of measurements (row length).

The smoothing parameter α is a weight coefficient for the actual level y_t and is selected by the researcher. In the case, when the row is stable and has small values of the random component ε_t, it is recommended to take values α close to one ($S_t \approx y_t$). In the case, where there is a pronounced influence ε_t on the value y_t, it is recommended to choose values α close to zero, since in this case, the maximum smoothing of the series levels is obtained ($S_t \approx S_{t-1}$).

In the simplest case, the value of S_0 is taken to be equal to the first of the levels of the series: $S_0 = y_1$. In order to improve the accuracy of the forecast, it is proposed to use the following expression to calculate S_0:

$$S_0 = y_1 - \frac{T_0}{2}, \tag{3.3}$$

where $T_0 = (y_N - y_1)/(N - 1)$.

To calculate the forecast value $\hat{y}_{t+1}$, use the expression:

$$\hat{y}_{t+1} = S_t \,. \tag{3.4}$$

It should be noted that for a time series with a pronounced trend, which is typical for most of the indicators analyzed in our study, standard exponential smoothing is ineffective, due to the fact that ignoring the trend dependence leads to the accumulation of a systematic error in the results. In this regard, a trend correction of the original model of standard exponential smoothing is required. Therefore, for the construction of predictive models, it is proposed to use the Holt model [8, 9] as one of the most effective models of exponential smoothing, taking into account the trend, in which the trend is also subjected to the smoothing procedure. It should be noted that the trend smoothing procedure is based on the recurrent ratio, which is similar to (3.2), but it uses a different personal parameter γ:

$$\begin{cases} S_t = \alpha\, y_t + (1-\alpha)(S_{t-1} + b_{t-1}); \\ b_t = \gamma(S_t - S_{t-1}) + (1-\gamma)\, b_{t-1}, \end{cases} \tag{3.5}$$

where t—time counts (in our case, years), $t = 1, 2, \ldots$;

y_t—the real level of the series at a given time t;

$\alpha.\ \gamma$—smoothing parameters ($0 < \alpha < 1$; $0 < \gamma < 1$);

S_t—smoothing value of real level y_t;

b_t—smoothed trend value for a point in time t.

When starting the computational process, it is necessary to set the initial values S_0 and b_0 in the recurrent Formulas (3.5) and choose the most effective smoothing parameters α and γ. At the initial stage, the trend value b_0 is usually chosen to be zero. The predicted values in the proposed Holt model are determined, based on the formula:

$$\hat{y}_{t+1} = S_t + b_t \,. \tag{3.6}$$

An important advantage of methods, based on the exponential smoothing model, is the possibility of additional consideration of the time value of information and adjustment, depending on changing conditions, which plays a significant role in the presence of unstable processes.

In accordance with the proposed methodology, predictive models were built for all the analyzed indicators.

When analyzing the personnel supply of the industry, the number of graduates in various specialties and areas of training was studied, divided by universities that train specialists for the needs of civil aviation, as well as the percentage of employment of university graduates.

Among the areas of training, the following were analyzed:

08.00.00 “Engineering and construction technologies” (secondary vocational education), including:

08.02.05 "Construction and operation of highways and airfields"—training is conducted only in the Yakut aviation technical school of civil aviation (a branch of the St. Petersburg state university of civil aviation);

11.00.00 "Electronics, Radio Engineering and Communication Systems" (Secondary Vocational Education), Including:

11.02.06 "Technical maintenance of transport radio-electronic equipment (air transport)"—training is conducted at the Rylsky aviation technical college (a branch of the Moscow state technical university of civil aviation), the Krasnoyarsk branch of the St. Petersburg state technical university of civil aviation, the Omsk flight technical college of civil aviation named after A. V. Lyapidevsky (a branch of the Ulyanovsk institute of civil aviation);

13.00.00 "Electric and Heat Power Engineering" (Secondary Vocational Education), Including:

13.02.11 "Technical operation and maintenance of electrical and electromechanical equipment"—training is conducted only in the Rylsky aviation technical college (a branch of the Moscow state technical university of civil aviation);

20.00.00 "Technosphere Safety and Environmental Management" (Higher Education), Including:

20.03.01 "Technosphere safety" (bachelor's degree)—training is conducted in all the main universities of civil aviation (Moscow state technical university of civil aviation, St. Petersburg state technical university of civil aviation, Ulyanovsk institute of civil aviation);

23.00.00 "Equipment and Technologies of Ground Transport" (Secondary Vocational and Higher Education), Including

programs SVE:

23.02.01 "Organization of transportation and management in transport (air transport)"—training is conducted at the St. Petersburg state technical university of civil aviation (in the main university and the three branches—Krasnoyarsk, Khabarovsk, Yakut aviation technical school of civil aviation);

23.02.03 "Maintenance and repair of air transport"—training is conducted in two branches of the Moscow state technical university of civil aviation (Yegoryevsky aviation technical college, Troitsky aviation technical college);

programs HE:

23.03.01 "Technology of transport processes" (bachelor's degree)—training is conducted in the Moscow state technical university of civil aviation and its Irkutsk branch, as well as in the St. Petersburg state technical university of civil aviation;

25.00.00 "Air Navigation and Maintenance of Aviation and Rocket and Space Technology" (Secondary Vocational and Higher Education), Including:

SVE programs:

25.02.01 "Technical maintenance of aircraft and engines"—training is conducted in five branches of the Moscow state technical university of civil aviation (Yegoryevsky aviation technical college, Irkutsk branch, Kirsanov aviation technical college, Rylsky aviation technical college, Troitsky aviation technical college), two branches of the St. Petersburg state technical university of civil aviation (Vyborg branch, Yakut aviation technical school of civil aviation) and the Omsk flight technical college of civil aviation named after A. V. Lyapidevsky (branch of the Ulyanovsk institute of civil aviation);

25.02.02 “Maintenance of aircraft with fuel and lubricants”—training is conducted at the Yegoryevsky aviation technical college (a branch of the Moscow state technical university of civil aviation) and the Krasnoyarsk branch of the St. Petersburg state technical university of civil aviation;

25.02.03 “Technical maintenance of electrified and flight-navigation complexes”—training is conducted in five branches of the Moscow state technical university of civil aviation (Yegoryevsky aviation technical college, Irkutsk branch, Kirsanov aviation technical college, Rylsky aviation technical college, Troitsky aviation technical college), two branches of the St. Petersburg state technical university of civil aviation (Vyborg branch, Yakut aviation technical school of civil aviation) and the Omsk flight technical college of civil aviation named after A. V. Lyapidevsky (branch of the Ulyanovsk institute of civil aviation);

25.02.04 “Flight operation of aircraft”—training is conducted in the St. Petersburg state technical university of civil aviation and its three branches (Krasnoyarsk branch, Yakut aviation technical school of civil aviation, Buguruslan flight school of civil aviation), as well as in three branches of the Ulyanovsk institute of civil aviation (Krasnokutsky flight school of civil aviation, Omsk flight technical college of civil aviation named after A. V. Lyapidevsky, Sasovsky flight school of civil aviation named after Hero of the Soviet Union Taran G. A.);

25.02.05 “Air traffic management”—training is conducted at St. Petersburg state technical University of civil aviation and its branches (Krasnoyarsk and Khabarovsk branches).

programs HE:

25.03.01 “Technical maintenance of aircraft and engines” (bachelor)—training is conducted in Moscow state technical university of civil aviation and its branches (Irkutsk and Rostov) and also at St. Petersburg state technical university of civil aviation and the central branch;

25.03.02 “Technical maintenance of aviation electrical systems and flight and navigation systems” (bachelor’s degree)—training is conducted at the Moscow state technical university of civil aviation and its two branches (Irkutsk and Rostov);

25.03.03 “Aeronavigation” (bachelor’s degree)—training is conducted in all the main universities (Moscow state technical university of civil aviation, St. Petersburg state technical university of civil aviation and Ulyanovsk institute of civil aviation);

25.03.04 “Airport operation and aircraft flight support” (bachelor’s degree)—training is conducted at the St. Petersburg state technical university of civil aviation and its Vyborg branch;

25.04.01 “Technical maintenance of aircraft and engines” (master’s degree)—training is conducted only in the Moscow state technical university of civil aviation;

25.04.02 “Technical maintenance of aviation electrical systems and flight and navigation systems” (master’s degree)—training is conducted only in the Moscow state technical university of civil aviation;

25.04.03 "Aeronavigation" (master's degree)—training is conducted at the St. Petersburg state technical university of civil aviation and the Ulyanovsk institute of civil aviation;

25.04.04 "Airport maintenance and aircraft flight support" (master's degree)—training is conducted only at the St. Petersburg state technical university of civil aviation;

25.05.03 "Technical maintenance of transport radio equipment" (specialty)—training is conducted at the Moscow state technical university of civil aviation and its two branches (Irkutsk and Rostov);

25.05.05 "Aircraft maintenance and air traffic management" (specialty)—training is conducted in all the main universities (Moscow state technical university of civil aviation, St. Petersburg state technical university of civil aviation and Ulyanovsk institute of civil aviation), as well as in the Vyborg branch of the St. Petersburg state technical university of civil aviation;

27.00.00 "Management in technical systems" (higher education), including:

27.03.02 "Quality management" (bachelor's degree)—training is conducted only at the Ulyanovsk institute of civil aviation;

43.00.00 Service and tourism (secondary vocational and higher education), including:

43.03.06 "Service in transport (by type of transport)" (bachelor's degree)—training is conducted only in the St. Petersburg state technical university of civil aviation;

43.02.06 "Service in transport (by type of transport)" (SVE program)—training is conducted in the Irkutsk branch of the Moscow state technical university of civil aviation, St. Petersburg state technical university of civil aviation.

As indicators of the development of air transport, the number of civil aviation airports, the number of flights (including domestic and international), passenger turnover (million pass. km). passenger traffic (million pass. km). cargo turnover (million ton km), the number of aircraft, served by airports, the number of passengers, served by airports, the number of cargo and mail passed through airports (tons) were analyzed. It should be noted that the last three indicators were analyzed both for each individual airport and for the Russian Federation as a whole, with a division into domestic, international and commercial flights.

In Tables 3.1 and 3.2, the dynamics of the number of graduates studying in technical specialties for the needs of civil aviation under the HE and SVE programs is presented.

As can be seen from the presented data, over the seven-year period from 2010 to 2016, both the HE and SVE programs show an increase in the total number of graduates in all areas of training for the needs of civil aviation (by 78.0% and 59.4%, respectively), in 2016, 4070 people were graduated from the HE programs, and 1757 people were graduated from the SVE programs.

In Fig. 3.1, the blue solid line represents the actual value, and the red dashed line represents the forecast.

In Fig. 3.2, the blue solid line represents the actual value, and the red dashed line represents the forecast.

Table 3.1 Dynamics of the number of graduates, studying in HE programs, according to data for the 2010–2016 years

Group of specialties and level of training	Year							Average value	Change in % to 2010 year
	2010	2011	2012	2013	2014	2015	2016		
Group 20.00.00	129	96	76	72	49	142	121	98	−6.2
Group 23.00.00	319	352	392	441	443	601	458	429	+ 43.6
Group 25.00.00	823	949	963	1235	1514	1498	1832	1259	+ 122.6
Group 27.00.00	46	45	70	96	94	28	41	60	−10.9
Group 43.00.00	54	69	61	81	61	45	43	59	−20.4
Bachelor's degree total	1287	1322	1295	1620	1834	2063	2090	1644	+ 62.4
Master's degree total	0	0	0	5	10	8	10	5	–
Specialty's degree total	155	189	267	300	317	243	395	267	+ 154.8
Total by HE	1442	1511	1562	1925	2161	2314	2495	1916	+ 73.0

Table 3.2 Dynamics of the number of graduates, studying in SVE programs, according to data for the 2010–2016 years

Group of specialties and level of training	Year							Average value	Change in % to 2010 year
	2010	2011	2012	2013	2014	2015	2016		
Group 08.00.00	0	17	18	0	10	17	13	11	−23.5
Group 11.00.00	137	106	121	164	129	156	153	138	+ 11.7
Group 13.00.00	40	23	23	27	28	31	24	28	−40.0
Group 23.00.00	315	248	195	246	181	198	191	225	−39.4
Group 25.00.00	718	980	1086	1268	1019	1614	1604	1184	+ 123.4
Group 43.00.00	161	175	199	145	162	155	200	171	+ 24.2
Total by SVE	1371	1549	1642	1850	1529	2171	2185	1757	+ 59.4

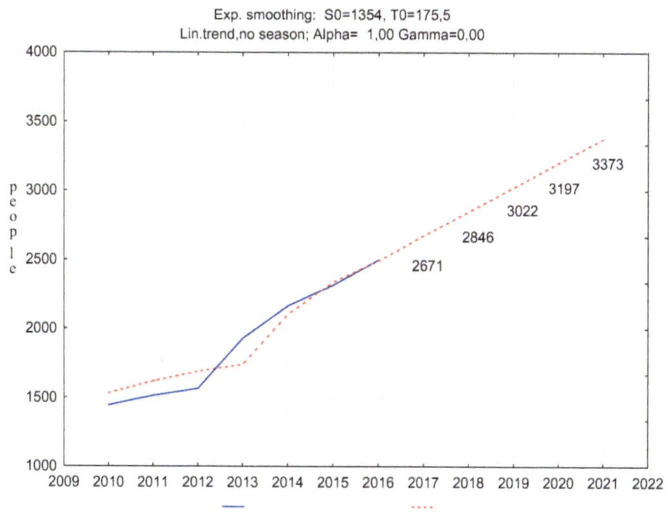

Fig. 3.1 Number of graduates in programs HE, based on data for 2010–2016 years with a forecast until 2022 year

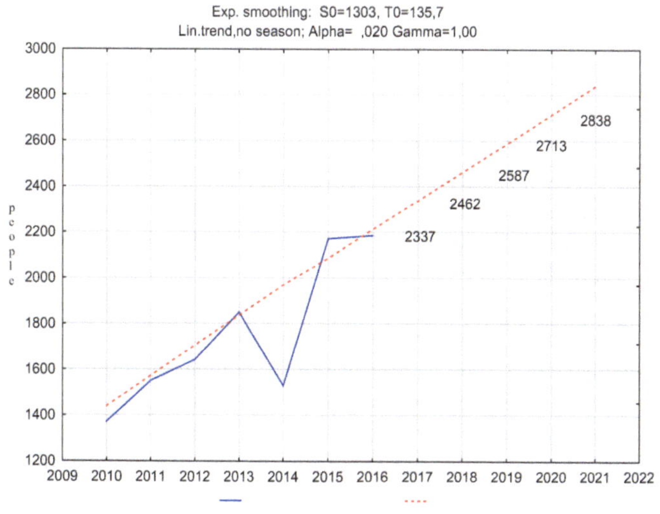

Fig. 3.2 The number of graduates in programs SVE, based on data for 2010–2016 years with a forecast until 2022 year

However, the situation is ambiguous in various areas of training. The largest increase in the graduation of specialists for the analyzed period, both in the HE and SVE programs, was observed in the group of specialties 25.00.00 (by 122.6% and 123.4%, respectively). In addition, among the HE programs, there was an increase in the number of graduates in the 23.00.00 specialty group by 43.6%, while in the

SVE programs in the same specialty group, there was a significant decrease in this indicator. For the other three groups of HE programs, there was a decrease in the number of graduates—by 20.4% in the group of specialties 43.00.00, by 10.9% in the group 27.00.00 and by 6.2% in the group 20.00.00. Among the SVE programs, the largest decrease in the number of graduates was observed in the group of specialties 13.00.00 (by 40.0%).

For all groups of specialties and in general for the HE and SVE programs (Figs. 3.1 and 3.2), short-term forecasts are based on the exponential smoothing method, according to which, if the current trend continues, the number of graduates in the HE and SVE programs should reach 3022 people and 2588 people by 2019, respectively. At the same time, an increase in the number of graduates in the following groups of specialties is predicted: 08.00.00 and 11.00.00 (SVE programs)—an average of 2 graduates per year; 23.00.00-HE programs (an average of 25 people per year); 25.00.00-HE programs (an average of 187 people per year) and SVE (an average of 144 people per year); 43.00.00—SVE programs (an average of 6 people per year). The decline in the number of graduates is predicted for the following groups of specialties: 13.00.00 (SVE programs), 22.00.00, 27.00.00 and 43.00.00 (HE programs)—an average of 2 graduates per year; 23.00.00 (SVE programs)—an average of 20 people per year.

It is also interesting to compare the number of graduates of the analyzed groups of specialties by university (Table 3.3). The largest increase in this indicator was registered in the Ulyanovsk institute of civil aviation—more than 2 times (an average of 16.9% per year).

The obtained results of short-term forecasting of the number of graduates in various areas of training allow us to take into account the existing general trends. However, one of the determining factors that must be taken into account. when predicting the number of graduates is the number of students, enrolled in the relevant time period (depending on the duration of training in the relevant programs—from 2 to 6 years before the projected moment).

In Table 3.4, the dynamics of the number of students enrolled in civil aviation universities is presented. From the presented data, it can be seen that MSTUCA and SPSUCA are characterized by an increase in the number of applicants for training, and the admission by SVE programs to SPSUCA for the analyzed period increased by more than one and a half times. For the UI CA, on the contrary, there is a tendency to a significant decrease in the number of persons, accepted under SVE programs—by 31.8% over 6 years.

From the point of view of forecasting the number of university graduates, it is interesting to analyze the share of the number of people, who have completed training in various groups of specialties in different universities and to study the existing dependencies. Table 3.5 and Figs. 3.3. 3.4 and 3.5 provide data on the share of graduates in relation to the number of students, enrolled in the analyzed educational programs, which shows that these figures differ for different universities. Thus, the maximum ratio of the number of graduates to the number of applicants to the university is typical for SPSUCA (70%) and the minimum—for MSTUCA (49%). It should be noted that for all universities, the ratio of the number of graduates

Table 3.3 Dynamics of the number of graduates of civil aviation universities according to the analyzed educational programs according to the data for 2010–2016 years

Higher education institution	Year							Average value	Change in % to 2010 year
	2010	2011	2012	2013	2014	2015	2016		
Moscow state technical university of civil aviation									
– Programs HE	454	418	408	568	557	635	839	554	+ 84.8
– Programs SVE	576	578	566	645	656	851	853	675	+ 48.1
– Total	1030	996	974	1213	1213	1486	1692	1229	+ 64.3
St. Petersburg state technical university of civil aviation									
– Programs HE	706	792	748	773	984	1053	1162	888	+ 64.6
– Programs SVE	628	675	744	813	580	873	861	739	+ 37.1
– Total	1334	1467	1492	1586	1564	1926	2023	1627	+ 51.6
Ulyanovsk institute of civil aviation									
– Programs HE	237	259	349	496	545	561	494	420	+ 108.4
– Programs SVE	167	279	314	392	283	430	320	312	+ 91.6
– Total	40 4	538	663	888	828	991	814	732	+ 101.5

Table 3.4 Dynamics of the number of persons entered in civil aviation universities according to the analyzed educational programs according to the data for 2010–2016 years

Higher education institution	Year							Average value	Change in % to 2010 year
	2010	2011	2012	2013	2014	2015	2016		
Moscow state technical university of civil aviation									
– Programs HE	787	884	883	873	854	918	981	883	+ 24.7
– Programs SVE	1274	1375	1270	1328	1351	1408	1524	1361	+ 19.6
– Total	2061	2259	2153	2201	2205	2326	2505	2244	+ 21.5
St. Petersburg state technical university of civil aviation									
– Programs HE	1180	1045	1128	1046	1233	1125	1203	1137	+ 1.9
– Programs SVE	973	1017	1006	1220	1183	1383	1466	1178	+ 50.7
– Total	2153	2062	2134	2266	2416	2508	2669	2315	+ 24.0
Ulyanovsk institute of civil aviation									
– Programs HE	637	654	682	623	722	746	734	685	+ 15.2
– Programs SVE	507	538	585	721	646	647	346	570	–31.8
– Total	1144	1192	1267	1344	1368	1393	1080	1255	–5.6

Table 3.5 Dynamics of the share of graduates in relation to the number of students, who entered in higher education institutions under the analyzed educational programs, according to data for 2010–2016 years

Higher education institution	Year							Average value	Change in % to 2010 year
	2010	2011	2012	2013	2014	2015	2016		
Moscow state technical university of civil aviation									
– Programs HE	0.58	0.47	0.46	0.65	0.65	0.69	0.86	0.62	+ 48.3
– Programs SVE	0.45	0.42	0.45	0.49	0.49	0.6	0.56	0.49	+ 24.4
– Total	0.5	0.44	0.45	0.55	0.55	0.64	0.68	0.54	+ 36.0
St. Petersburg state technical university of civil aviation									
– Programs HE	0.6	0.76	0.66	0.74	0.8	0.94	0.97	0.78	+ 61.7
– Programs SVE	0.65	0.66	0.74	0.67	0.49	0.63	0.59	0.63	–9.2
– Total	0.62	0.71	0.7	0.7	0.65	0.77	0.76	0.7	+ 22.6
Ulyanovsk institute of civil aviation									
– Programs HE	0.37	0.4	0.51	0.8	0.75	0.75	0.67	0.61	+ 81.1
– Programs SVE	0.33	0.52	0.54	0.54	0.44	0.66	0.92	0.56	+ 178.8
– Total	0.35	0.45	0.52	0.66	0.61	0.71	0.75	0.58	+ 114.3

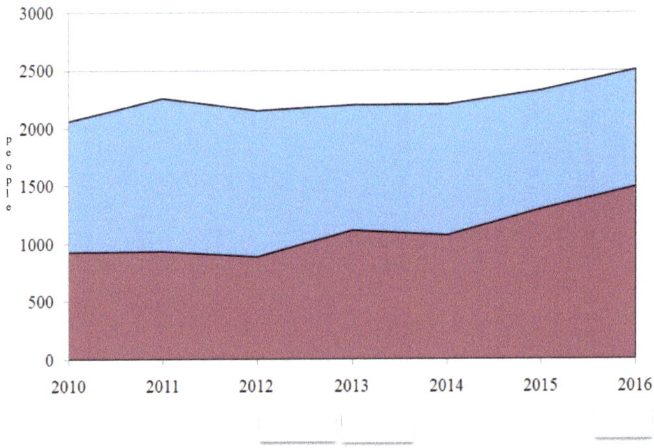

Fig. 3.3 Dynamics of the share of graduated specialists in relation to the number of MSTUCA applicants, according to data for 2010–2016 years

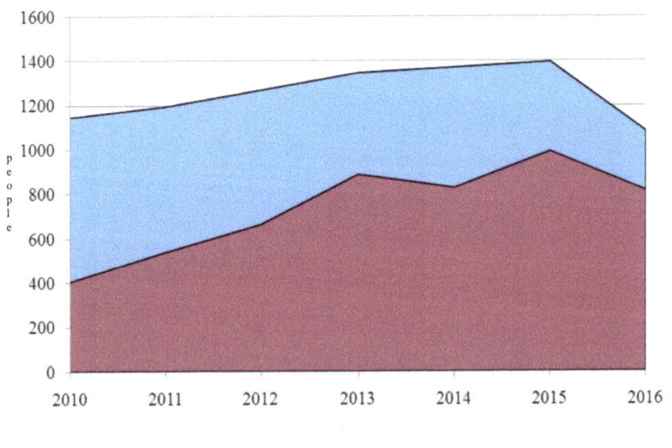

Fig. 3.4 Dynamics of the share of graduated specialists in relation to the number of SPSUCA applicants, according to data for 2010–2016

to the number of applicants for HE programs is higher than for SVE programs—the difference is from 5% (UI CA) to 22% (MSTUCA). Similar differences are typical for different groups of specialties (Tables 3.6 and 3.7), which should be taken into account, when making a forecast. At the same time, over the past 6 years, there has been a tendency to increase the share of graduates (an increase of up to 60%).

The blue color indicates admission to university, and the maroon color indicates the graduation of students.

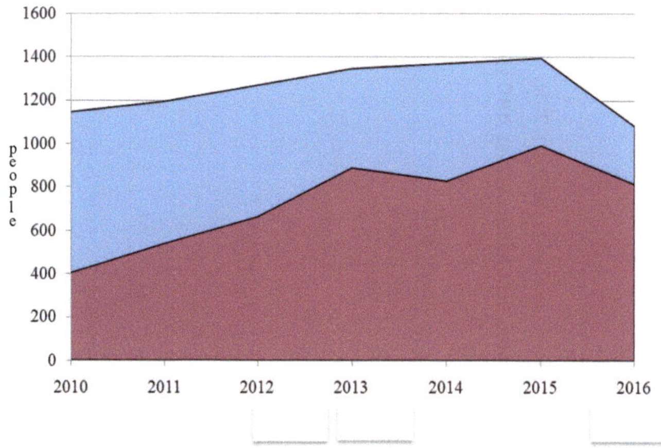

Fig. 3.5 Dynamics of the share of graduated specialists in relation to the number of UI CA applicants, according to data for 2010–2016

The blue color indicates admission to university, and the maroon color indicates the graduation of students.

The blue color indicates admission to institute, and the maroon color indicates the graduation of students.

As can be seen from the presented data, there are various complex relationships between the number of students, who entered and graduated from various educational programs and training levels, and therefore, mathematical models, describing the existing relationships, were built to obtain a forecast.

In accordance with the methodology, described in Chap. 2, several alternative models were built for each indicator. A total of 25 different models were used, and the full list of which is given in Table 3.8.

As an example, Table 3.9 shows alternative models for predicting the number of graduates in bachelor's degree programs.

The optimal model was selected, based on statistical criteria and expert assessments in accordance with the methodology, proposed in Chap. 2. Table 3.10 presents models for predicting the number of graduates in training programs in various civil aviation universities, and Fig. 3.6 presents an example of a graphical description of the identified dependencies.

When evaluating the prospective values of the volume indicators of the graduation of specialists within the educational cluster of civil aviation, it is necessary to focus not only on the effectiveness of the activities of educational organizations, but also on the projected performance indicators of air companies, taking into account the development of air transport. To solve this problem, the dynamics of a number of leading indicators of civil aviation airports were analyzed, and the corresponding predictive models were constructed using the Holt model, described above.

Table 3.11 shows the main analyzed indicators with generalized data for the Russian Federation. Figures 3.7 and 3.8 and Table 3.12 show the results of forecasting

Table 3.6 Dynamics of the share of graduates in relation to the number of applicants for the analyzed HE programs, according to data for 2010–2016

Group of specialties and level of training	Year							Average value	Change in % to 2010 year
	2010	2011	2012	2013	2014	2015	2016		
Group 20.00.00	40.7	61.9	48.7	61.5	52.1	169	147.6	83	+ 262.7
Group 23.00.00	57.6	78.9	78.1	95	100.9	160.3	91.2	95	+ 58.3
Group 25.00.00	42.5	48	46.8	61.8	65.2	66.3	77.7	58	+ 82.8
Group 27.00.00	79.3	95.7	127.3	320	324.1	121.7	178.3	178	+ 124.8
Group 43.00.00	67.5	123.2	203.3	300	762.5	900	268.8	375	+ 298.2
Bachelor's degree total	52.4	85.8	78	102.9	101.3	127.8	105.7	93	+ 101.7
Master's degree total	0	0	0	100	90.9	9.1	9.4	30	–
Specialty's degree total	54	16.7	23.6	28.4	29.6	23.2	44	31	−18.5
Total by HE	52.6	56.3	55.8	73	74.7	84.2	83.7	69	+ 59.1

Table 3.7 Dynamics of the share of graduates in relation to the number of applicants for the analyzed HE programs, according to data for 2010–2016

Group of specialties and level of training	Year							Average value	Change in % to 2010 year
	2010	2011	2012	2013	2014	2015	2016		
Group 08.00.00	0	48.6	51.4	0	33.3	56.7	39.4	33	−18.9
Group 11.00.00	34.3	28.3	43.7	52.6	45.7	54	48.7	44	+ 42.0
Group 13.00.00	181.8	109.5	164.3	81.8	–	91.2	61.5	115	−66.2
Group 23.00.00	114.1	98	80.9	103.4	46.8	61.3	50.8	79	−55.5
Group 25.00.00	38.5	47	51.8	51.4	45.2	65	63	52	+ 63.6
Group 43.00.00	83.9	87.9	85.8	66.5	62.8	50.3	59.9	71	−28.6
Total by SVE	49.8	52.2	56.7	56.2	47.6	62.6	60	55	+ 20.5

Table 3.8 Types of used models

#	Model name	General view of the model
1	Linear	$Y = a + b * X$
2	Square root-Y	$Y = (a + b * X)^2$
3	Exponential	$Y = \exp(a + b * X)$
4	Reciprocal-Y	$Y = 1/(0.0.138.137 + 0.000.112.727 * X)$
5	Squared-Y	$Y = (a + b * X)^{1/2}$
6	Square root-X	$Y = a + b * \sqrt{X}$
7	Double square root	$Y = (a + b * \sqrt{X})^2$
8	Logarithmic-Y square root-X	$Y = \exp(a + b * \sqrt{X})$
9	Reciprocal-Y square root-X	$Y = 1/(0.01.047 + 0.00.127.985 * \sqrt{X})$
10	Squared-Y square root-X	$Y = (a + b * \sqrt{X})^{1/2}$
11	Logarithmic-X	$Y = a + b * \ln X$
12	Square root-Y logarithmic-X	$Y = (a + b * \ln X)^2$
13	Multiplicative	$Y = \exp(a + b * \ln X)$
14	Reciprocal-Y logarithmic-X	$Y = 1/(a - b * \ln X)$
15	Squared-Y logarithmic-X	$Y = (a + b * \ln X)^{1/2}$
16	Reciprocal-X	$Y = a - b/X$
17	Square root-Y reciprocal-X	$Y = (a - b/X)^2$
18	S-curve model	$Y = \exp(a - b/X)$
19	Double reciprocal	$Y = 1/(a + b/X)$
20	Squared-Y reciprocal-X	$Y = (a - b/X)^{1/2}$
21	Squared-X	$Y = a + b * X^2$
22	Square root-Y squared-X	$Y = (a + b * X^2)^2$
23	Logarithmic-Y squared-X	$Y = \exp(a + b * X^2)$
24	Reciprocal-Y squared-X	$Y = 1/(a - b * X^2)$
25	Double squared	$Y = (a + b * X^2)^{1/2}$

Note ***a*** and ***b*** are the parameters of model (coefficients), determined using the least squares method (LSM)

individual generalized indicators of airport activity (in the Russian Federation as a whole). Similar forecasts were made for all the analyzed indicators, both in the Russian Federation as a whole and at the level of individual airports.

According to the calculations, a slight decrease in the number of airlines (on average by 2 air company per year), general aviation operators (by 6 per year), an increase in the number of aircraft (by 9.6% annually) and the number of operators performing aviation work is projected. According to all indicators that characterize the activities of civil aviation, their growth is projected: the number of flights—by

Table 3.9 Alternative models for predicting the number of graduates in bachelor's degree programs

Model name	General view of the model
Linear	$Y = 2.98111 + 0.603816 * X$
Square root-Y	$Y = (2.94256 + 0.050656 * X)^2$
Exponential	$Y = \exp(2.32968 + 0.019036 * X)$
Squared-Y	$Y = (-1146.89 + 53.9679 * X)^{1/2}$
Square root-X	$Y = -24.7106 + 8.58526 * \sqrt{X}$
Double square root	$Y = (0.431574 + 0.748225 * \sqrt{X})^2$
Logarithmic-Y square root-X	$Y = \exp(1.28661 + 0.295992 * \sqrt{X})$
Squared-Y square root-X	$Y = (-3386.34 + 732.243 * \sqrt{X})^{1/2}$
Logarithmic-X	$Y = -59.351 + 24.9447 * \ln X$
Square root-Y logarithmic-X	$Y = (-3.01295 + 2.28902 * \ln X)^2$
Multiplicative	$Y = \exp(-0.308723 + 0.968423 * \ln X)$
Reciprocal-Y logarithmic-X	$Y = 1/(0.325191-0.736115 * \ln X)$
Squared-Y logarithmic-X	$Y = (-5852.98 + 1995.67 * \ln X)^{1/2}$
Reciprocal-X	$Y = 43.3937-317.945/X$
Square root-Y reciprocal-X	$Y = (6.55536-33.4306/X)^2$
S-curve model	$Y = \exp(3.81961-16.5805/X)$
Double reciprocal	$\mathbf{Y = 1/(-0.00274336 + 1.68958/X)}$
Squared-Y reciprocal-X	$Y = (2226.94-21183.5/X)^{1/2}$
Squared-X	$Y = 17.6361 + 0.00450529 * X^2$
Square root-Y squared-X	$Y = (4.23131 + 0.000360493 * X^2)^2$
Logarithmic-Y squared-X	$Y = \exp(2.84266 + 0.000127017*X^2)$
Reciprocal-Y squared-X	$Y = 1/(0.0736677-0.00000612448 * X^2)$
Double squared	$Y = (80.2752 + 0.427034 * X^2)^{1/2}$

Table 3.10 Models for predicting the number of graduates depending on the number of people, accepted for study

Training level	Type of model	Model
MSTUCA		
Secondary vocational training	Multiplicative	$Y = \exp(0.462742 + 0.748281 * \ln X)$
Bachelor's degree	Double squared	$Y = (-271.75 + 0.551798 * X^2)^{1/2}$
Master's degree	Square root-Y squared-X	$Y = (1.90182 + 0.00869053 * X^2)^2$
Specialty's degree	Double squared	$Y = (211.901 + 0.441921 * X^2)^{1/2}$
SPSUCA		
Secondary vocational training	Squared-X	$Y = 14.0245 + 0.00403816 * X^2$
Bachelor's degree	Double reciprocal	$Y = 1/(-0.00126424 + 1.68443/X)$
Master's degree	Square root-Y squared-X	$Y = (1.90182 + 0.00869053 * X^2)^2$
Specialty's degree	Double squared	$Y = (211.901 + 0.441921 * X^2)^{1/2}$
UI CA		
Secondary vocational training	Double squared	$Y = (-1753.72 + 0.684671 * X^2)^{1/2}$
Bachelor's degree	Double squared	$Y = (136.026 + 0.484622 * X^2)^{1/2}$
Master's degree	Square root-Y squared-X	$Y = (1.90182 + 0.00869053 * X^2)^2$
Specialty's degree	Double squared	$Y = (211.901 + 0.441921 * X^2)^{1/2}$

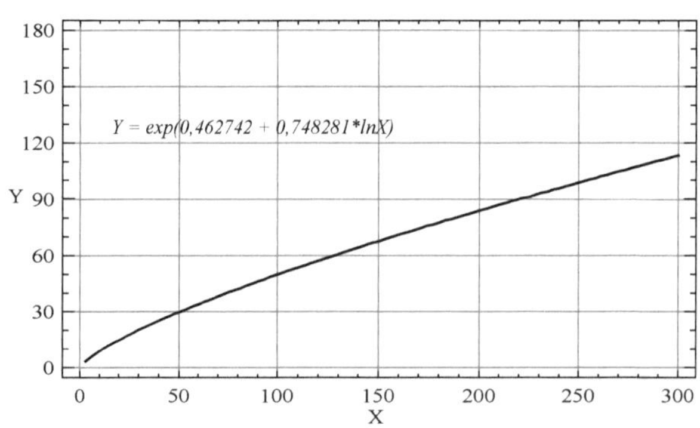

Fig. 3.6 Model for predicting the number of graduates, depending on the number of students, entered in MSTUCA by programs SVE

Table 3.11 Dynamics of the main indicators of the functioning of civil aviation, as a whole in the Russian Federation

Indicator name	Year							Average value	Change in % to 2010 year
	2010	2011	2012	2013	2014	2015	2016		
Number of air companies	127	127	121	118	121	115	113	120	−11.0
Number of operators, general aviation	98	99	73	74	68	72	66	79	−32.7
Number of operators, aviation works	190	216	211	224	227	240	251	223	+ 32.1
Number of aircrafts, total	1219	1452	1815	2076	2400	2670	2876	2073	+ 135.9
Number of flights, total	1,109,663	1,248,106	1,318,486	1,418,749	1,462,131	1,427,270	1,413,319	1,342,532	+ 27.4
Number of flights, domestic	432,784	480,135	496,352	547,087	619,811	683,424	716,866	568,066	+ 65.6
Number of flights, international	676,879	767,971	822,134	871,662	842,320	743,846	696,453	774,466	+ 2.9
Passenger turnover, billion passenger km	147.1	166.78	195.7	225.16	241.42	226.8	215.6	202.65	+ 46.6
Passenger traffic, million passengers	56.95	64.12	74.00	84.56	93.2	92.07	88.56	79.07	+ 55.5

(continued)

Table 3.11 (continued)

Indicator name	Year							Average value	Change in % to 2010 year
	2010	2011	2012	2013	2014	2015	2016		
Cargo turnover, million tons km	4715	4950	5076	5011	5151	5481	6575	5280	+ 39.4
Cargo and mail transportation, thousand tons	926	981	988	1001	1036	971	977	983	+ 5.5
Total number of airports	234	238	238	246	254	255	255	246	+ 9.0
Number of aircrafts, served by airports, piece	2,655,252	2,950,104	3,351,736	3,565,949	3,779,472	3,744,012	3,632,866	3,382,770	+ 36.8
Number of passengers, served by airports, people	184,671,635	208,130,404	253,441,782	284,781,388	315,503,474	318,660,322	319,196,091	269,197,871	+ 72.8
Quantity of cargo and mail passed through airports, tons	1,463,281	1,745,111	2,003,476	1,888,326	1,761,330	1,538,311	1,680,392	1,725,747	+ 14.8

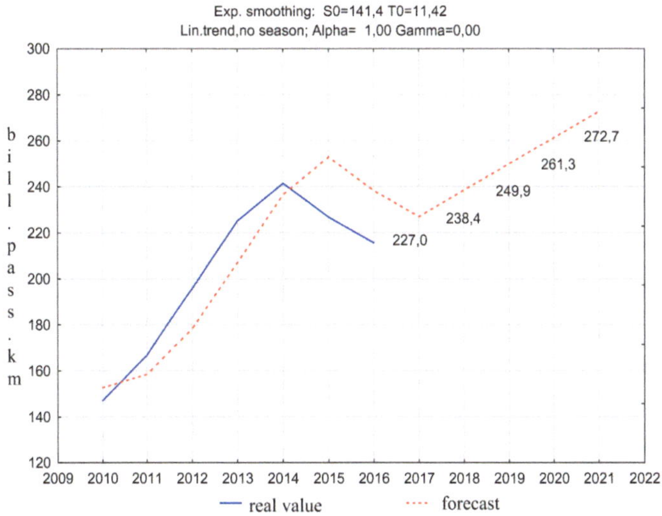

Fig. 3.7 Passenger turnover of Russian airports (billion passenger-kilometers), according to data for 2010–2016 with a forecast until 2022

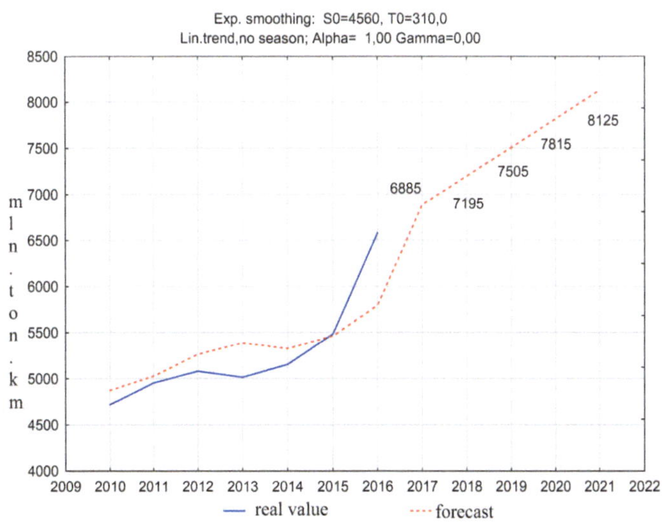

Fig. 3.8 Cargo turnover of airports of the Russian Federation (million tons km) according to data for 2010–2016 with a forecast until 2022

3.6% annually (with the largest increase in the number of domestic flights with an almost constant number of international flights); passenger turnover—by 5.3%, and passenger traffic—by 6.0%; cargo turnover—by 4.7% (at the same time, the volume of transported goods should increase only by 1.0% annually). The number of airports

Table 3.12 Results of short-term forecasting of the main indicators of the functioning of civil aviation (in the whole of the Russian Federation)

Indicator name	Year					Average annual change
	2017	2018	2019	2020	2021	
Number of operators, general aviation	111	108	106	104	102	−2 (−2.1%)
Number of operators, aviation works	58	53	47	42	37	−6 (−9.6%)
Number of aircrafts, total	263	272	281	291	300	+ 10 (+ 4.0%)
Number of flights, total	1,463,928	1,514,538	1,565,147	1,615,756	1,666,366	+ 50609 (+ 3.6%)
Number of flights, domestic	764,213	811,560	858,907	906,254	953,601	+ 47347 (+ 6.6%)
Number of flights, international	699,715	702,978	706,240	709,502	712,765	+ 3262 (+ 0.5%)
Passenger turnover, billion passenger km	227	238.4	249.9	261.3	272.7	+ 11.4 (+ 5.3%)
Passenger traffic, million passengers	93.8	99.1	104.4	109.6	114.9	+ 5.3 (+ 6.0%)
Cargo turnover, million tons km	6885	7195	7505	7815	8125	+ 310 (+ 4.7%)
Cargo and mail transportation, thousand tons	989.8	998.3	1006.8	1015.3	1023.8	+ 10 (+ 1.0%)
Total number of airports	260	264	267	271	274	+ 4 (+ 1.6%)
Number of aircrafts, served by airports, piece	3,795,802	3,958,737	4,121,673	4,284,609	4,447,544	+ 162936 (+ 4.5%)
Number of passengers, served by airports, people	352,827,205	375,247,948	397,668,690	420,089,433	442,510,176	+ 261575533 (+ 8.2%)
Quantity of cargo and mail passed through airports, tons	1,716,577.2	1,752,762.3	1,788,947.5	1,825,132.7	1,861,317.8	+ 36185 (+ 2.2%)

in the Russian Federation is expected to increase by an average of 4 (1.6%) annually, At the same time, the number of aircrafts, served by airports, should increase by 4.5% annually, the number of passengers—by 8.2%, and the number of cargo and mail passed through airports—by 2.2%.

Similar forecasts (for 5 years ahead—until 2022) were built for 255 civil aviation airports, located on the territory of the Russian Federation. We analyzed such indicators as the number of served aircrafts (with the allocation of international, domestic and commercial flights, taking into account regular and irregular flights), the number of passengers, cargo and mail, which were sent. received and in transit. According to the received forecasts, in 213 (83.5%) airports, a further increase in the number of served aircraft is projected. An increase in the number of served passengers is projected at 219 airports (85.9%), an increase in the volume of received and sent cargo is projected at 153 airports (60.0%), and an increase in the volume of received and sent mail is projected at 149 airports (58.4%).

Thus, using the proposed multi-method approach, predictive models are constructed for all the analyzed indicators. The obtained short-term forecasts should be taken into account, when planning the long-term values of the volume indicators of graduates of civil aviation universities. An important role in this study is played by the use of GIS-oriented information about the geographically distributed system of airports, universities, with reference to the analyzed indicators.

With the use of geoinformation systems, it becomes possible to generate spatial and temporal information, based on statistical data, linked, on the one hand, to calendar periods, and on the other, to the geolocation of airports in the Russian Federation and civil aviation universities (Fig. 3.9). At the same time. one part of the information support characterizes the efficiency of air transport by a number of indicators: cargo turnover, passenger turnover, passenger traffic volume, the number of passengers, served by airports, the degree of renewal of the aircraft fleet; the second is the personnel training system: indicators for monitoring the performance of industry-specific educational organizations and other information systems of the Ministry of education and science of the Russian Federation.

Based on GIS-based information, it becomes possible to estimate the volume of graduation and employment of graduates of industry-specific educational organizations, depending on the location of airports. In Fig. 3.10, the spatially-temporal data on the number of people, accepted for training and graduated from civil aviation universities with reference to the geolocation of airports, are presented. In Fig. 3.11, projected dynamics (based on the results of short-term forecasting) of the number of graduates with reference to the projected change in airport performance indicators is presented. Similar spatially distributed data were analyzed for all areas of training of civil aviation universities (Fig. 3.12). Based on this analysis. it is possible to identify those air companies. for which it is necessary to optimize the mechanisms for managing their staffing support.

The results of predictive modeling provide an opportunity for expert evaluation of the prospective values of the volume indicators of the personnel training system in the area of air transport maintenance.

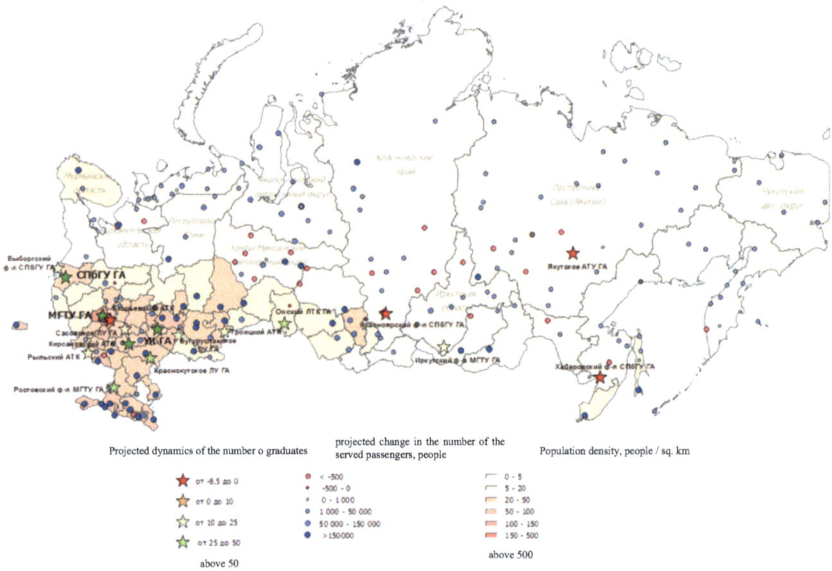

Fig. 3.9 Geolocation of airports of the Russian Federation and civil aviation universities

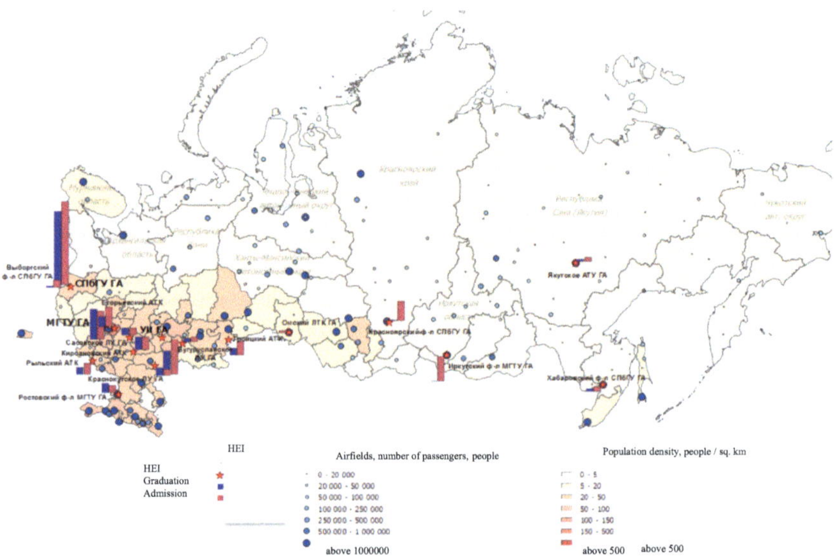

Fig. 3.10 Spatially-temporal data about the number of people, accepted for study and graduated from civil aviation universities and the number of the served passengers by airports of the Russian Federation, according to data for 2016

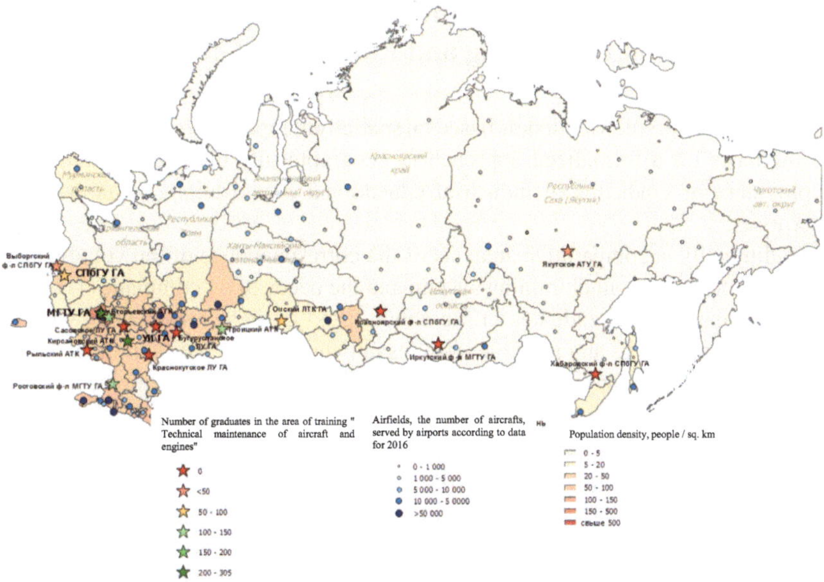

Fig. 3.11 Projected dynamics (change for the period from 2016 to 2019 year) of the number of graduates with reference to the projected change in the performance of airports, according to data for 2016

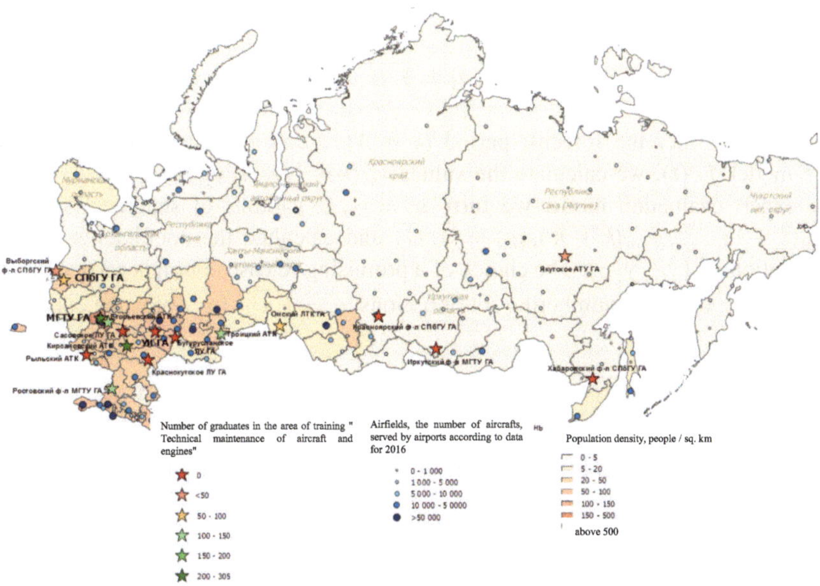

Fig. 3.12 Data about the number of graduates in the area of training "technical maintenance of aircraft and engines" and the number of aircraft, served by airports, according to data for 2016

3.2 Expert and Statistical Estimation Perspective the Amount of Training and Employing Them

The constructed predictive models based on spatially-temporal information resources are the basis for a formalized approach to geographically distributed planning of employment of graduates of industrial educational organizations in air transport enterprises.

Traditionally, a comparative analysis of the correspondence of the volume characteristics of the personnel training system and the needs of air companies is carried out by comparing the graduates in the above-listed areas and specialties with the vacancies of personnel and possible changes in their number. This approach defines a one-sided forecast for a short planning period but does not take into account the trends in the development of the industry in terms of efficiency.

In addition to the predictive model $f_g(t)$. an important characteristic of the adequacy of staffing support to the needs of air companies is the degree of coherence between the efficiency indicators of the functioning and development of air transport and the volume indicators of graduates in areas and specialties.

To quantify the degree of connectivity, we define the numbering set $g_1 = \overline{1, G_1} \in 1, G$, for key industry indicators y_{g_1}, and the numbering set $g_2 = \overline{1, G_2} \in 1, G$ for volume graduate indicators y_{g_2}. From the values of these indicators, recorded at time points t_k, we will form statistical samples $y_{g_1}(k)$, $y_{g_2}(k)$for$k = 1, 2 \ldots$ time intervals. Using methods of processing of statistical data, we determine the coefficients of pair correlation for k time intervals

$$\rho^k(y_{g_1}, y_{g_2}), g_1 = \overline{1, G_1}, g_2 = \overline{1, G_2}.$$

Next, consider the forecast period $(k + 1)$, ..., $(k + K)$. Using the predictive model $f_{g_1}(t)$, we calculate the values $\hat{y}_{g_1}(k + 1), \ldots, \hat{y}_{g_1}(k), g_1 = \overline{1, G_1}$. In the expert evaluation mode. we form $u = \overline{1, U}$ variants of statistical samples $\hat{y}_{g_2u}(k + 1), \ldots, \hat{y}_{g_2u}(k + K), g_2 = \overline{1, G_2}$ and calculate the coefficients of pair correlation $\rho_u^K(\hat{y}_{g_1}, \hat{y}_{g_2u})$. The choice of a promising release option is carried out as follows: We enter the numbering set of options $u_1 = \overline{1, U_1}$. for which the condition is met

$$\rho_{u_1}^K(\hat{y}_{g_1}, \hat{y}_{g_2u}) > \rho^k(y_{g_1}, y_{g_2}); \tag{3.7}$$

we select $y_{g_2}^*$, $g_2 = \overline{1, G_2}$ from variants $u_1 = \overline{1, U_1}$. which meets the condition

$$\max_{u_1} \rho_{u_1}^K(\hat{y}_{g_1}, \hat{y}_{g_2u_1}). \tag{3.8}$$

Thus, the estimates $\hat{y}^{*}_{g_2}(k+1), \ldots, \hat{y}^{*}_{g_2}(k+K), g_2 = \overline{1, G_2}$ should be considered as the total needs of the industry for personnel with professional and higher education, on the basis of which the control figures of admission (CFA) for educational organizations are established.

The process of forecasting is finished by geographically distributed planning of employment of graduates of avia transport enterprises on the basis of GIS-related information (Fig. 3.13).

In the framework of the proposed approach, based on the model of Holt, described in the previous section, it was built the predictions of the analyzed indicators of efficiency of functioning and development of air transport for the five-year period ($K = 5$) from 2017 to 2021 year. At the same time, forecasting was carried out at two levels:

(1) at the level of data for the Russian Federation as a whole (Table 3.12)—to determine the prospective need for specialists and determine the control figures of admission in general for groups of specialties and individual areas of training;
(2) at the level of individual air transport enterprises (appendix with reference to their geographical location—for the rational distribution of the allocated CFA limit between individual universities and branches, taking into account their territorial location).

For the indicators that characterize the efficiency of the functioning and development of air transport in general at the level of the Russian Federation, the coefficients of pair correlation were calculated, which characterize the relationship of the analyzed indicators with the volume of graduation of specialists with higher and secondary vocational education in various groups of specialties and areas of training. At the same time, in accordance with the proposed approach, real data for the period from 2010 to 2016 and projected prospective values for 2017–2021 were analyzed. Taking into account the fact, that specialists, who have completed their studies at the university and got a job, contribute to the activities of the civil aviation enterprise in full in the year following the year of graduation, when calculating the correlation coefficients, the offset of the analyzed data for 1 year was taken into account, that is, data on the effectiveness of the activities of air enterprises for 2011 were compared with the data on the graduation of specialists in 2010, etc. A fragment of the calculations is presented in Table 3.13. The analysis of the relationship of the predicted values was carried out to form the initial options for consideration, when using expert procedures—if the relationship of the predicted indicators is statistically significant and meets the condition (3.7), then the forecast can be accepted by the expert as one of the options u_1.

However, not all correlation coefficients are of interest for planning the long-term need for specialists, but only those ones that characterize the relationship between the analyzed indicators of the effectiveness of the functioning and development of air transport with the graduation of specialists in areas, that provide the corresponding type of activity (Table 3.14).

Let us consider the work of the proposed algorithm for forming a plan for the admission and graduation of specialists in general for all HE and SVE programs

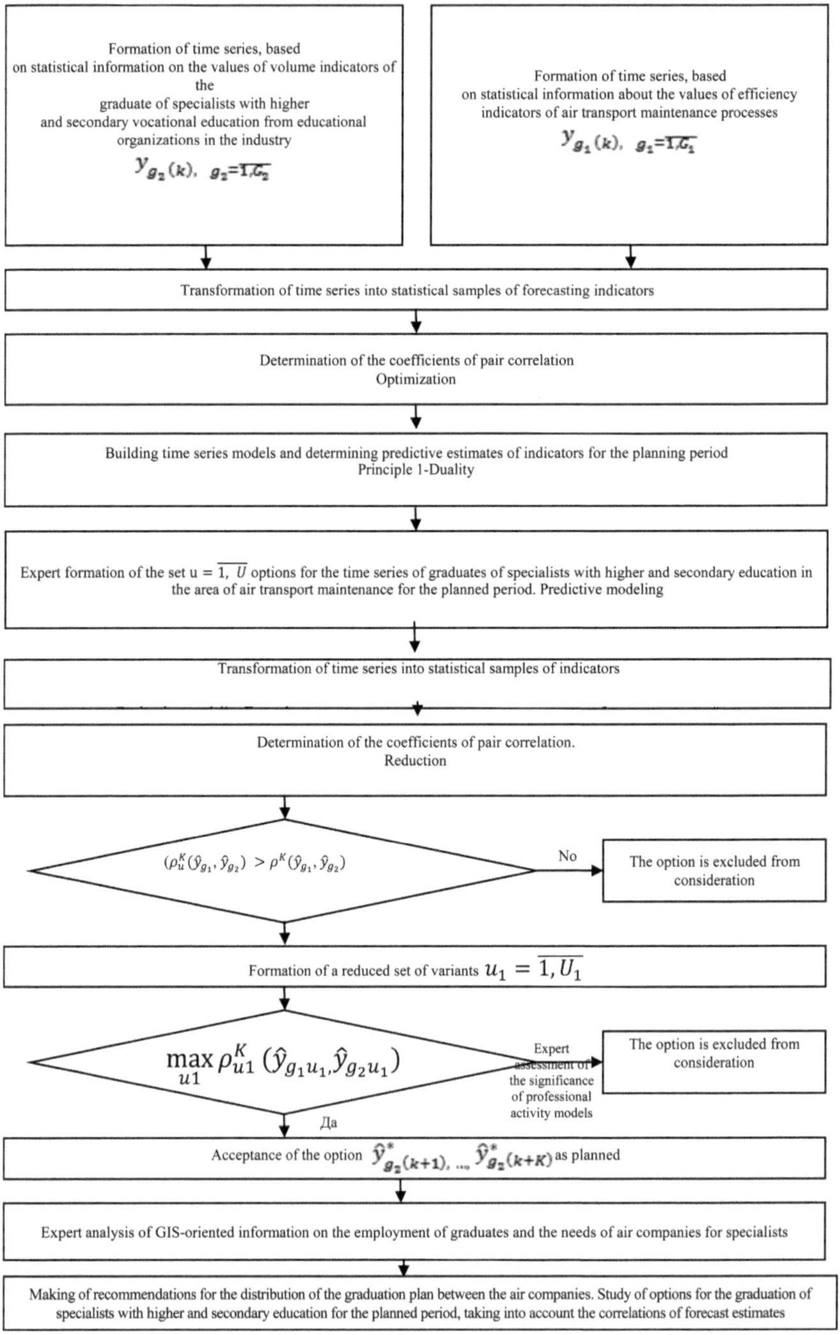

Fig. 3.13 Structural scheme of personnel training planning in the area of air transport maintenance

Table 3.13 Relationship (coefficients of pair correlation) of the performance indicators of the functioning and development of air transport with the volume of graduation of specialists in various groups of specialties and areas of training, according to data for 2010–2016

Code of the training area	Indicator of the functioning and development of air transport							
	Number of air companies	Number of flights	Passengers turnover	Cargo turnover	Total number of airports	Number of aircraft. served by airports	Number of passengers. served by airports	The number of cargo and mail that passed through the airports
20.03.01	−0.18	−0.35	−0.37	0.31	0.02	−0.27	−0.13	**−0.77**
23.03.01	**−0.81**	**0.76**	**0.76**	0.48	**0.83**	**0.82**	**0.83**	−0.14
25.03.01	**−0.85**	0.64	0.59	**0.95**	**0.82**	0.65	**0.77**	−0.12
25.03.02	−0.39	0.47	0.52	0.31	0.45	0.43	0.46	0.15
25.03.03	−0.75	0.44	0.40	**0.83**	**0.76**	0.51	0.64	−0.50
25.03.04	0.23	0.34	0.41	−0.43	0.12	0.27	0.16	0.38
25.05.03	−0.50	0.50	0.53	0.48	0.63	0.47	0.55	−0.17
25.05.05	**−0.78**	**0.77**	0.72	**0.77**	0.66	**0.78**	**0.79**	0.51
27.03.02	0.07	0.35	0.42	−0.34	0.01	0.29	0.17	0.66
43.03.06	0.41	−0.02	−0.02	−0.62	−0.40	−0.15	−0.29	0.60
08.02.05	−0.20	0.26	0.14	0.34	0.21	0.27	0.24	0.22
11.02.06	−0.74	0.45	0.50	0.41	0.55	0.49	0.57	−0.22
13.02.11	0.30	−0.50	−0.36	−0.38	−0.22	−0.42	−0.35	**−0.76**
23.02.01	**0.79**	**−0.86**	**−0.83**	−0.70	**−0.87**	**−0.90**	**−0.91**	−0.18
23.02.03	−0.70	0.46	0.47	0.69	**0.81**	0.53	0.67	−0.62
25.02.01	**−0.86**	0.66	0.57	**0.97**	**0.76**	0.66	0.75	0.05
25.02.02	0.69	**−0.78**	**−0.78**	−0.67	**−0.81**	**−0.83**	**−0.86**	−0.19
25.02.03	**−0.95**	**0.76**	0.73	**0.81**	**0.90**	**0.81**	**0.89**	−0.14
25.02.04	**−0.90**	0.74	0.68	0.64	0.72	**0.77**	**0.79**	0.12
25.02.05	**−0.89**	0.61	0.54	**0.78**	0.74	0.62	0.71	−0.17
43.02.06	−0.13	−0.07	−0.16	0.52	−0.08	−0.04	−0.00	0.32

Note The correlation coefficients that characterize the reliable relationships for $p < 0.05$ are highlighted in bold

Table 3.14 Indicators of the effectiveness of the functioning and development of air transport and the areas of specialist training that provide them

Indicator	Areas of training
Number of avia companies	27.03.02, 43.03.06, 23.02.01, 25.02.04, 25.05.05, 27.03.02, 43.03.06, 23.02.01
Number of operators, general aviation	27.03.02, 43.03.06, 23.02.01, 25.02.04, 25.03.01, 25.05.05, 27.03.02, 43.03.06, 23.02.01
Number of operators, avia works	20.03.01, 25.03.01, 27.03.02, 43.03.06, 13.02.11, 25.02.01, 25.02.02
Number of aircrafts, total	23.03.01, 25.03.01, 25.03.02, 25.03.03, 25.05.03, 25.05.05, 08.02.05, 11.02.06, 13.02.11, 25.02.01, 25.02.02, 25.02.04
Number of flights, total	23.03.01, 25.03.01, 25.03.02, 25.03.03, 25.03.04, 25.05.05, 25.02.03, 25.02.04
Number of flights, domestic	23.03.01, 25.03.01, 25.03.02, 25.03.03, 25.03.04, 25.05.05, 25.02.05
Number of flights, international	25.03.01, 25.03.02, 25.03.04, 25.05.05, 27.03.02, 25.02.04, 25.02.05
Passenger turnover, billion pass. km	23.03.01, 25.03.02, 25.05.03, 11.02.06,
Passenger traffic, million pass	08.02.05, 23.02.01, 43.02.06
Cargo turnover, million t. km	20.03.01, 23.02.01, 25.02.01
Cargo and mail transportation, thousand tons	25.03.03, 25.03.04, 23.02.01, 23.02.03
Total number of airports	23.03.01, 25.03.01, 25.03.02, 25.03.03, 25.03.04, 25.05.03, 25.05.05, 27.03.02, 43.03.06, 08.02.05, 11.02.06, 23.02.03, 25.02.03, 25.02.05, 43.02.06
Number of aircraft, served by airports, piece	20.03.01, 23.03.01, 25.03.01, 25.03.02, 25.03.03, 25.03.04, 25.05.03, 25.05.05, 11.02.06, 13.02.11, 25.02.01, 25.02.02, 25.02.03, 25.02.04, 25.02.05
Number of passengers, served by airports, people	25.03.04, 27.03.02, 43.03.06, 08.02.05, 23.02.01, 23.02.03, 43.02.06
Quantity of cargo and mail passed through airports, tons	25.03.04, 43.03.06, 23.02.01

(the numbering set g_2 includes two elements ($G_2 = 2$), corresponding to the volume of graduation in the bachelor's degree and SVE areas). The key industry indicators (set g_1) for solving this problem are "total number of airports". "Number of aircraft, served by airports", "number of passengers, served by airports" and "number of cargo and mail, passed through airports" ($G_1 = 4$). Dynamics of indicators for 2010–2016 years and their prognostic estimates for the period from 2017 to 2021 years are presented in Tables 3.1, 3.2, 3.11 and 3.12.

The coefficients of pair correlation $\rho^k(y_{g1}.y_{g2})$, calculated on the basis of statistical information on the values of volume indicators of the graduation of specialists with higher and secondary vocational education from educational organizations in the industry, and indicators of the efficiency of air transport maintenance processes, are presented in Table 3.15.

The results of the expert formation of a set options of time series ($u = \overline{1, U}$) for the graduation of specialists with higher and secondary education in the area of air transport maintenance for the planned period (5 years) are shown in Table 3.16.

Table 3.17 shows the results of calculations of the pair correlation coefficients $\left(\rho_u^K(\hat{y}_{g_1} \cdot \hat{y}_{g_2}).u = \overline{1.U}\right)$ between the expert variants of the volumes of graduation of specialists with higher and secondary vocational education in 2017–2021 years

Table 3.15 Relationship between the volume of graduation of specialists in the HE and SVE programs and the performance indicators of air transport maintenance, according to data for 2010–2016 years

Efficiency indicators of air transport maintenance processes	The volume of the graduation of specialists	
	By programs HE	By programs SVE
Total number of airports	**0.98**	0.74
Number of aircraft, served by airports	**0.85**	0.67
Number of passengers, served by airports	**0.93**	0.75
The number of cargo and mail that passed through the airports	−0.15	−0.07

Note The correlation coefficients that characterize reliable relationships for $p < 0.05$ are highlighted in bold

Table 3.16 Results of expert formation of a set of variants of time series of the volume of graduation of specialists

Variant number	Year				
	2017	2018	2019	2020	2021
HE programs					
Variant 1	2400	2440	2360	2480	2720
Variant 2	2215	2430	2310	2505	2690
Variant 3	2220	2390	2410	2570	2720
Variant 4	2600	2590	2580	2670	2760
Variant 5	2250	2410	2360	2690	2650
SVE programs					
Variant 1	2230	2455	2370	2605	2600
Variant 2	2150	2260	2355	2450	2550
Variant 3	2210	2320	2300	2520	2570
Variant 4	2000	2280	2220	2340	2420
Variant 5	2215	2350	2350	2510	2510

Table 3.17 Relationship of expert options for the graduation of specialists with higher and secondary vocational education in 2017–2021 years with the predicted values of the efficiency indicators of air transport maintenance processes

Efficiency indicators of air transport maintenance processes	The volume of the specialist's graduation				
	Variant 1	Variant 2	Variant 3	Variant 4	Variant 5
HE programs					
1. Total number of airports	0.75	0.89	0.98	0.83	0.91
2. Number of aircraft, served by airports	0.76	0.89	0.98	0.83	0.90
3. Number of passengers, served by airports	0.76	0.89	0.98	0.83	0.90
4. The number of cargo and mail that passed through the airports	0.76	0.89	0.98	0.83	0.90
SVE programs					
1. Total number of airports	0.91	0.99	0.96	0.91	0.96
2. Number of aircraft, served by airports	0.88	0.99	0.95	0.89	0.95
3. Number of passengers, served by airports	0.88	0.99	0.95	0.89	0.95
4. The number of cargo and mail that passed through the airports	0.88	0.99	0.95	0.89	0.95

and the predicted values of the efficiency indicators of air transport maintenance processes.

As a result of comparing the coefficients, characterizing the relationship of the series of analyzed data for the period from 2010 to 2016 years with the coefficients of the pair correlation of the predicted values, options 1, 2, 4, 5 for the HE programs were excluded from further consideration as not corresponding to the condition $\rho_u^K(\hat{y}_{g_1}, \hat{y}_{g_2}) > \rho^k(y_{g_1}, y_{g_2})$.

Thus. the reduced set of variants ($u_1 = \overline{1.U_1}$) for the HE programs included only Variant 3 which was accepted as planned. For SVE programs, all the variants, proposed by the experts, were included in the set $u_1 = \overline{1.U_1}$ and when using the proposed selection rule $\left(\max\limits_{u1} \rho_{u1}^K(\hat{y}_{g_1u_1} \cdot \hat{y}_{g_2u_1})\right)$. Variant 2 was chosen as the planned one.

Naturally, the determining factor, affecting the number of specialists, who have completed education is the number of applicants, accepted for education. In this connection, to ensure the implementation of the long−term plan for the graduation of specialists, it is necessary to form an appropriate admission plan, which is also associated with control figures of admission (CFA) for budget places.

The formation of an adapted admission plan is proposed to be carried out on the basis of the developed predictive models (Tables 3.9 and 3.10), describing the relationship between the number of persons, who have completed education in various

Table 3.18 Planned values of figures of admission for higher and secondary special education programs for 2017–2018 years

Level of education program	Year	
	2017	2018
Education programs HE	3060	3240
Education programs SVE	3950	4110

programs and the number of applicants for education (in the corresponding year—depending on the level of the educational program and the form of education). As a result of the calculations, the planned values of the control figures admission for higher and secondary special education programs for 2017–2018 years were determined (Table 3.18).

Similar calculations for determining the plan of graduation and admission of specialists were performed for each individual area of training in higher and secondary education programs.

To solve the problem of the distribution of graduates among Russian air companies, it is proposed to base on the results of expert analysis of GIS–oriented information [10, 11] about the needs of airlines for specialists. We will base on the fact that the need of air companies for specialists of various profiles depends on the predicted dynamics of the analyzed indicators of the efficiency of air transport maintenance processes (Table 3.12). With negative dynamics of indicators, there is no need for new specialists; with positive dynamics, the need for specialists of the corresponding profile (Table 3.14) depends on the projected increase in the corresponding indicator. At the same time, the territorial mutual location of educational institutions and air companies is taken into account. using GIS technologies. An example of the distribution plan for 2018 year graduates studying by the higher professional education program in the enlarged group of areas 25.00.00 "air navigation and maintenance of aviation and rocket and space technology" is shown in Fig. 3.14.

Thus, the combination of statistical data processing of time series of indicators of the efficiency of air transport maintenance processes and volume indicators of graduation from various HE and SVE programs from educational organizations, based on the determination of the coefficients of pair correlation and forecast options, proposed by the results of expert evaluation, ensures the feasibility of management decisions on planning the volume of graduates and their employment in a geographically distributed system of civil aviation.

3.3 Expert Assessment and Selection of Priority Types of Professional Activity in the Framework of Their Multi-stage Reduction

The analysis of the dynamics of key indicators allows not only to take into account the trends in the volume indicators of personnel training in accordance with the needs of airlines, but also to allocate educational resources that ensure the formation of

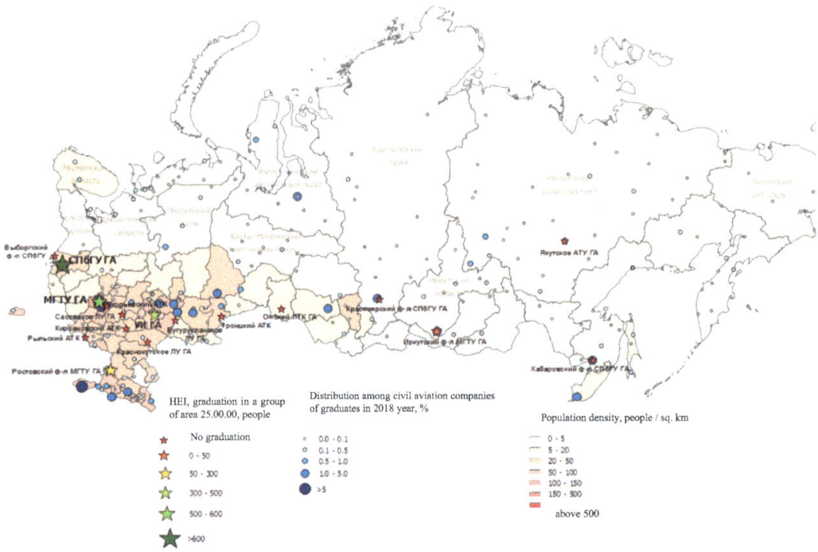

Fig. 3.14 Planned distribution among civil aviation companies of graduates of 2018 year, studying in an enlarged group of areas 25.00.00

knowledge, skills and abilities for professional activity in the most popular areas of industry development [10, 11].

Long–term planning of the personnel training system is associated not only with the choice of volume indicators. but also with the definition of priority activities of air transport maintenance specialists, focused on the industry development strategy, To solve the second problem, which is the first step in the implementation of multi-stage reduction, it is proposed to implement evaluation modeling, based on expert information.

10 different types of professional activities of air transport maintenance specialists were analyzed:

(1) development of plans, programs and methods of work during the maintenance of aircraft;
(2) conducting research to reduce the loss of material resources. labor and time in the process of technical maintenance of aircraft;
(3) improving the efficiency of air transport enterprise management by key performance indicators;
(4) ensuring and improving the efficiency of technical maintenance of aircraft; operational reliability, regularity of flights; intensity and efficiency of use;
(5) development of airport and airfield infrastructure;
(6) operation and maintenance of aircraft;
(7) provision and maintenance of air transportation and aviation works;
(8) organization, execution, provision and maintenance of aircraft flights;

(9) organization of maintenance and repair of aircraft and aviation infrastructure facilities;
(10) organization and provision of services for passengers, baggage, cargo and mail.

The group examination involved 4 specialists from the universities of the industry and 4 specialists from the management bodies of the air companies. The method of a priori ranking was used. The experts were asked to organize (rank) the types of professional activities, depending on the degree of their significance for the development of the industry. Table 3.19 shows the results of a survey of 8 experts.

Taking into account that some experts have assigned the same rank to different types of professional activities (related ranks have appeared), to perform further calculations, it is necessary to switch to the normal (standardized) form of the ranking matrix, in which the values of related ranks are defined as the sum of places, divided among themselves by alternatives with the same ranks. The results of the calculations are presented in Table 3.20.

The value of the concordance coefficient is 0.86. The calculated value of the χ^2-criterion of Pearson $\chi^2{}_{\text{расч}} = 62{,}277$ is greater than the 5% limit for the Pearson distribution with the number of degrees of freedom (9) $\chi^2{}_{\text{кр}} = 16{,}919$. Therefore, the hypothesis of the consistency of experts is accepted.

Based on the generalized ranks $r_m, m = \overline{1, M}$, we construct a ranking histogram (Fig. 3.15). From the visual analysis of the uneven distribution, it follows that the most significant areas are

oving the efficiency of management an air transport enterprise by key performance indicators ($r_3 = 1$);

development of airport and airfield infrastructure ($r_5 = 2$).

The priority of these areas is explained by the following reasons:

- increasing the importance of transportation on domestic lines;
- the need to strengthen the manageability and coordination of the activities of the civil complex;

Table 3.19 Matrix of ranking of professional activities types

Activity types numbers	1	2	3	4	5	6	7	8	9	10
Number $m = \overline{1, M}$ of expert $d = \overline{1, D}$										
I	9	10	1	3	2	5	7	4	6	8
II	6	6	1	3	2	3	4	3	5	4
III	10	9	2	5	1	3	8	4	6	7
IV	7	8	1	3	2	3	5	4	6	5
V	3	7	2	4	1	4	6	5	3	6
VI	9	10	1	4	2	5	6	3	7	8
VII	7	7	1	2	1	3	5	4	6	5
VIII	7	8	1	2	1	2	4	3	5	6

Table 3.20 Computational matrix for ranking of the activity types

Activity types numbers	1	2	3	4	5	6	7	8	9	10
Number $m = \overline{1, M}$ of the expert $d = \overline{1, D}$										
I	9	10	1	3	2	5	7	4	6	8
II	9.5	9.5	1	3	2	4.5	6.5	4.5	8	6.5
III	10	9	2	5	1	3	8	4	6	7
IV	9	10	1	3.5	2	3.5	6.5	5	8	6.5
V	3.5	10	2	5.5	1	5.5	8.5	7	3.5	8.5
VI	9	10	1	4	2	5	6	3	7	8
VII	9.5	9.5	1.5	3	1.5	4	6.5	5	8	6.5
VIII	9	10	1.5	3.5	1.5	3.5	6	5	7	8
$\sum_{d=1}^{D} a_{dm}$	68.5	78	11	30.5	13	34	55	37.5	53.5	59
b	24.5	34	−33	−13.5	−31	−10	11	−6.5	9.5	15
b^2	600.25	1156	1089	182.25	961	100	121	42.25	90.25	225
Ranks z_m	9	10	1	3	2	5	8	4	6	7

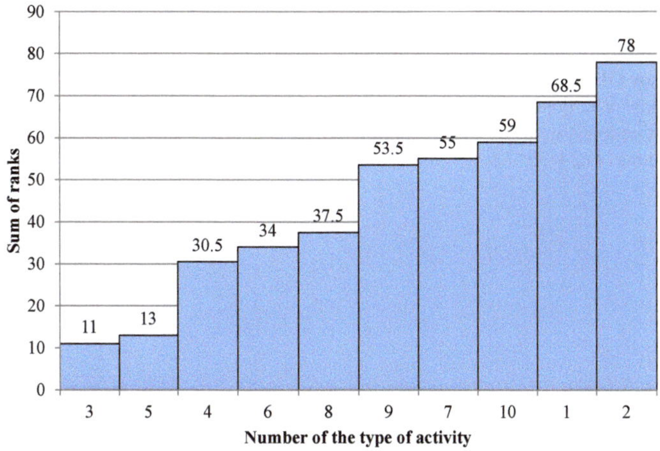

Fig. 3.15 Ranking histogram (Diagram of the ranks of the types of professional activities of specialists in the maintenance of air transport)

- implementation of measures to ensure the availability of services for the population;
- increasing the volume of work on the construction and reconstruction of the airport and airfield infrastructure.

Thus. as part of the implementation of the model-oriented approach, based on the multi-stage reduction procedure, at the first stage, on the basis of evaluation modeling and expert selection of priority professional activities, taking into account the projected graduation volumes in the areas and specialties of professional and higher education in the area of air transport maintenance, the choice of priority professional types of activity of specialists is justified.

References

1. Marienkin EV, Eliseev BP, Elisov LN (2011) About the concept of system engineering management by aviation staffing of civil aviation personnel. Sci Bull MSTUCA (174):43–47
2. Meltzer MI (1986) Dialog management of production. In: Finance and statistics, p 240
3. Khalafyan AA (2008) Statistica 6. Statistical analysis of data, 3rd edn. Textbook. Binom–Press, p 512
4. Nolan Richard L (1979) Managing the crises in data processing. Harvard Bus Rev 57(2):115–116
5. Andersson K, Hall W, Atkins S, Feron E (2003) Optimization–based analysis of collaborative airport arrival planning. Transp Sci 422–433
6. Simplifying Passenger Travel website. http://www.spt.aero
7. Simplifying the Business website. http://www.iata.org/whatwedo/stb
8. Vukolov EA (2008) Fundamentals of statistical analysis. In: Workshop on statistical methods and operations research using STATISTICS and EXCEL packages: textbook—M: FORUM, p 464

9. Gaskarov DV, Golinkevich TA, Mozgalevsky AV (1974) Forecasting the technical condition and reliability of radio−electronic equipment. M Sov Radio, p 224
10. Kuznetsov OL, Nikitin AA (1992) Geoinformation. M Nedra, p 302
11. Mikryukov VN, Ponevazh VP, Seregin AN, Sorokin SO (2015) Research of opportunities of geoinformation systems for the analysis of activity of educational organizations. Reform Differ Dimensions (1):34–39

Chapter 4
Optimization of Personnel Training Based on the Use of Invariant Models of Professional Activity in the Field of Air Transport Enterprise Management

The previous chapter shows that one of the essential activities in the area of air transport maintenance, for which it is advisable to assess the possibility of a problem orientation of the formalized apparatus of the model-oriented approach, is to increase the efficiency of management of an air transport enterprise. To this end, first of all, it is necessary to continue the procedure of multi-stage reduction, based on expert information, to rank a variety of areas in the field of management to determine the most promising. The reduced subset of directions requires further detailing to the level of certain competencies and labor functions of specialists, associated with the management of an air transport enterprise. The performance of labor functions requires knowledge and skills, focused on the use of certain model components. The final stage of reduction is aimed at determining invariant mathematical constructions, the content, and parametric identification of which allows us to build a content description of each model of professional activity that supports the implementation of detailed labor functions of the selected promising direction in the area of air company management.

The presence of invariant descriptions of professional activity models makes it possible to distribute more effectively the planned labor intensity during the transition to training elements by transforming them into components of educational programs.

4.1 Expert Selection of Promising Areas of Training in the Field of Air Company Management and Their Corresponding Invariant Models

Practical aspects of personnel training in the area of air company management, related to the formation of labor functions and competencies, corresponding to this type of activity, require the inclusion of educational resources in a number of areas in the content of study in the area of air transport maintenance [1, 2], given in Table 4.1.

A. Borzova et al., *Modeling and Forecasting of Staffing in Civil Aviation*, Springer Aerospace Technology, https://doi.org/10.1007/978-981-19-6238-7_4

Table 4.1 Main directions of personnel training in the area of enterprise management

Number in order	Name of the direction
1	Anti-crisis management
2	Risk management and insurance
3	Quality system management
4	Innovative management
5	Management of foreign economic activity and international relations
6	Management of production activities
7	Financial and economic activity management

The complexity of training in this area should be divided between the given areas in accordance with their significance from positions that are ahead of the requirements for ensuring an effective process of air transport maintenance. In the case of a model-oriented approach, it is necessary to organize a multi-stage reduction of the components of professional activity in the following sequence:

(1) reduction of the original set of directions;
(2) reduction of professional activity models, corresponding to the directions chosen at the first stage;
(3) reduction of invariant descriptions of professional activity models, selected at the second stage.

At the first stage of reduction, we use the method of group expert evaluation—a priori ranking [3]. 8 experts participate in the survey. They are invited to arrange the directions, given in Table 4.1, according to the degree of decreasing importance for training personnel in a certain educational area or specialty in the area of air transport maintenance. Table 4.2 shows the results of a survey of 8 experts from among the leading specialists of aviation enterprises and universities.

Table 4.2 Matrix of expert survey

Number of expert эксперта $d = \overline{1,\ D}$	Number of the directions $m = \overline{1,\ M}$						
	1	2	3	4	5	6	7
I	4	6	2	5	7	1	3
II	5	1	3	7	4	2	6
III	3	4	1	6	7	2	5
IV	4	3	2	7	6	1	5
V	5	6	2	4	7	1	3
VI	6	7	1	3	4	2	5
VII	2	5	4	7	6	1	3
VIII	5	6	1	7	3	2	4

Table 4.3 Computational matrix

d	m						
	1	2	3	4	5	6	7
I	4	6	2	5	7	1	3
II	5	1	3	7	4	2	6
III	3	4	1	6	7	2	5
IV	4	3	2	7	6	1	5
V	5	6	2	4	7	1	3
VI	6	7	1	3	4	2	5
VII	2	5	4	7	6	1	3
VIII	5	6	1	7	3	2	4
$\sum_{d=1}^{D} a_{dm}$	34	38	16	46	44	12	34
b	2	6	−16	14	12	−20	2
b^2	4	36	256	196	144	400	4
ranks z^m	3,4	5	2	7	6	1	3,4

Let us move from the matrix of expert survey to the computational matrix (Table 4.3).

The value of the concordance coefficient is 0.44. The calculated value of the Pearson criterion $\xi^2_{\text{расч}}$=29,36 is greater than the 5% limit for the Pearson distribution with the number of degrees of freedom $\xi^2_{\kappa p} = 12{,}592$. Therefore, the hypothesis about the consistency of experts is accepted.

Having estimated the consistency of experts by the concordance coefficient and having determined the ranks r_m, $m = \overline{1, M}$, we will construct a ranking histogram (Fig. 4.1). From the visual analysis of the uneven distribution, it follows that the most significant areas are:

management of production activities ($r_6 = 1$);

quality system management ($r_3 = 2$).

Let us move on to the second stage of reduction at the level of professional activity models. To characterize the management of production activities, the following models are needed [1, 4]:

(1) selection of key indicators of activity of production activities;
(2) determining the frequency of monitoring of indicators;
(3) forecasting of monitoring of the primary characteristics of the production process;
(4) calculating the values of key indicators, based on the monitoring results;
(5) probabilistic assessment of evidence of deviation of the actual values of indicators from the planned ones;
(6) probabilistic assessment of the causes of deviations;
(7) making management decisions to compensate for deviations.

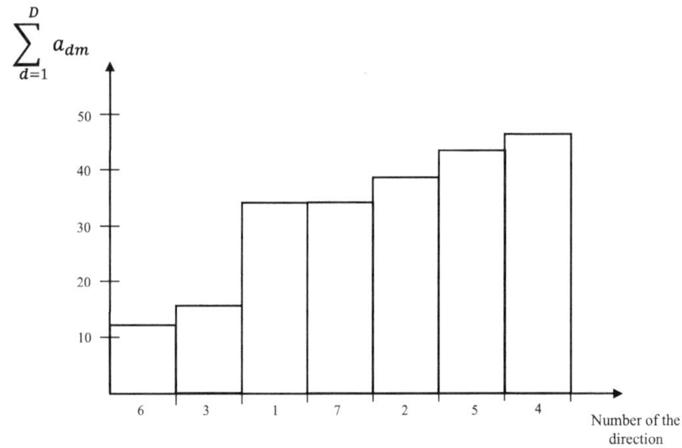

Fig. 4.1 Ranking histogram (diagram of the rank direction in the area of managing the production activities of an air company)

Quality system management is characterized by the following models of professional activity [5–10]:

(1) selection of criteria for the quality of air transport products;
(2) determining the frequency of monitoring criteria;
(3) monitoring and forecasting of the primary quality characteristics of air transport products;
(4) calculating the values of quality criteria, based on the monitoring results;
(5) probabilistic assessment of evidence of non-compliance with the quality standard;
(6) probabilistic assessment of the causes of non-compliance;
(7) selection of corrective actions to ensure compliance with the quality standard.

We use the same ranking method as for the first stage of reduction. As a result, we get that the most significant in terms of the complexity of the study are the models of professional activity, given in Table 4.4. The same table shows the results of the third stage, associated with the definition of invariant mathematical descriptions of professional activity models, selected at the second stage.

Let us focus on invariant descriptions of models, related to the management of production activities. Similarly, a description of the models, that characterize the management of the quality system, is formed. They can be extended to other types of management.

Thus, for anti-crisis management, the above invariant mathematical constructions serve as the basis for describing the following models of professional activity, related to this area:

Table 4.4 Invariant mathematical descriptions of models of professional activity in the area of air company management

The name of the professional activity models, selected at the 2nd stage		The type of invariant mathematical description
Production activity management quality system management	Quality system management	
Selection of key indicators of activity of production activities	Selection of quality criteria for air transport products	Optimization model
Probabilistic assessment of the reasons for deviations of the actual values of key indicators from the planned ones	Probabilistic assessment of the reasons for non-compliance with the quality standard	Inductive logic model
Making management decisions to compensate of deviations	Selection of corrective actions to ensure compliance with the quality standard	Optimization model

- selection of indicators of the occurrence of pre-crisis and crisis situations;
- probabilistic assessment of the causes of pre-crisis situations;
- making management decisions to get out of pre-crisis situations.

4.2 Optimization Modeling of the Selection of Key Indicators of Activity in the Tasks of Air Company Management

Optimization modeling as an invariant content of the basic tasks of managing the maintenance of air transport within the air company consists of three components:

- a meaningful description of the management task;
- formation of an optimization model;
- construction of a procedure that allows you to find a solution to an optimization problem.

The balanced scorecard methodology is used as the basis for the meaningful formulation of the optimal choice problem, which provides the possibility of coordinating various levels of company management, based on key indicators of activity (KIA) [11].

KIA is intended for the company's management as a decision-making tool in the process of managing the company's activities. These can be decisions, related to the development of the company's activities or to the reorganization of the company's activities, including the improvement of business processes, the improvement of organizational structures, the introduction of information technologies, etc. (Fig. 4.2).

Such use of KIA requires the formation of business goals, critical success factors, key indicators of activity and measures (actions) to achieve goals. Business goals are defined as the priority strategic directions of the company's business development,

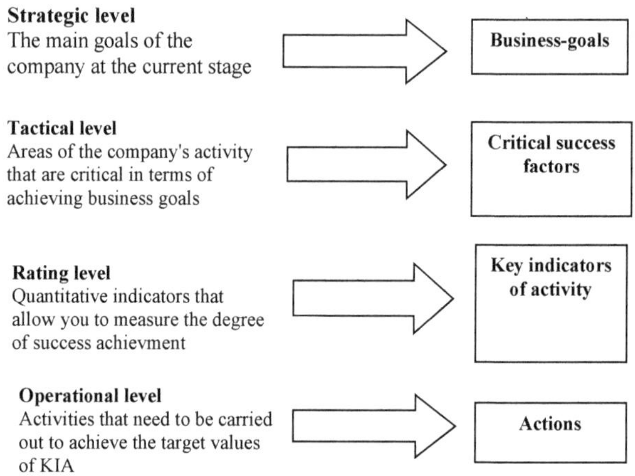

Fig. 4.2 Relationship of KIA with different levels of management

which determine the conditions for the company's successful existence on the market in the medium and long term.

The management system by deviations reacts to change in KIA values in case of their deviation of the threshold value by launching a certain set of pre-determined measures, recorded in the corresponding plans or instructions.

The control loop by deviations will include the following processes:

- measurement of KIA values;
- analysis of the criticality of detected deviations;
- informing about the detected deviation;
- making a decision to launch corrective procedures (for example, issuing an order for the company to put into effect a pre-prepared action plan).

The positive dynamics of the indicators indicates the presence of positive trends in the external and internal environments, which may be the basis for revising KIA's target values, and in cases of serious success—for changing strategic priorities at the level of the company's business goals.

The informative statement serves as the basis for the formation of an optimization model for choosing the KIA of the control problem by deviations. The complete set of indicators of the efficiency of the air transport maintenance process $y_g, g = \overline{1, G}$ and control (corrective) actions $x_j, j = \overline{1, J}$ is considered. Next, an expert ranking of the set of control actions x_j is carried out, characterizing their significance for compensating of the deviation of the actual values of the g-th, $g = \overline{1, G}$ indicator from the planned ones, with the determination of the value of the ranks r_{jg}, varying in the interval $\left[\overline{1, J}\right]$. According to the rank diagram r_{jg}, similar to the one, shown in Fig. 4.1, two rank situations are visually selected, in which jumps of the generalized sum of the ranks of the jth impact occur. This allows us to divide the control actions into 3 groups, according to the level of significance of the influence on reducing the deviation of the gth indicator: significant, average, insignificant.

The results of the expert evaluation are formalized, using a discrete value c_{jg}, $j = \overline{1,\ J}, g = \overline{1,\ G}$:

$$c_{jg} = \begin{cases} 1, & \text{if the level of significance is essential;} \\ 0.5, & \text{if the level of significance is mean;} \\ 0, & \text{if th level of significance is nonessential.} \end{cases} \quad (4.1)$$

Based on a meaningful description of the KIA selection task, the number of these indicators should be the minimum possible. Let us introduce an alternative variable

$$z_g = \begin{cases} 1, & \begin{array}{l}\text{if } g\text{th the efficiency indicator of the air transport maintenance process} \\ \text{is included in the number of KIA for solving control problems,}\end{array} \\ 0, & \text{in otherwise, } g = \overline{1, G}. \end{cases} \quad (4.2)$$

Then, the optimization criterion that provides the requirement of minimization has the form

$$\sum_{g=1}^{G} z_g \to \min. \quad (4.3)$$

The boundary conditions are related to the requirement, that at least one control action, that has a essential level of significance, corresponds to each gth indicator. These conditions are formalized, using the following system of constraints

$$\sum_{g=1}^{G} c_{jg} z_g \geq 1, \quad j = \overline{1, J}. \quad (4.4)$$

Let us combine the optimization criterion (4.3) and constraints (4.2), (4.4) into a single optimization model:

$$\begin{gathered} \sum_{g=1}^{G} z_g \to \min, \\ \sum_{g=1}^{G} c_{jg} z_g \geq 1, \quad j = \overline{1,\ J}, \\ z_g = \begin{cases} 1, \\ , \\ 0 \end{cases} \quad g = \overline{1,\ G}. \end{gathered} \quad (4.5)$$

The final stage of optimization modeling is the construction of a procedure for solving the problem (4.5). The solution is based on the scheme of randomized dichotomous reduction, considered in clause 2.1 in the form of an iterative algorithm. At the first step of the iterative process, we assume a uniform distribution of randomized variables z_g, $g = \overline{1, G}$.

$$p_{z_g}^1 = 0,5, \quad q_{g_g}^1 = 0,5. \tag{4.6}$$

To change the distribution (4.6) in following iterations:

$$p_{z_g}^{k+1} = p_{z_g}^k + A\left(F\left(\tilde{z}_g^k\right)\right) \tag{4.7}$$

is necessary first of all to determine the values of some equivalent function for the problem (4.5), and on its basis to build a predictive estimate that determines the direction of movement by changing the distributions of variables $z_{g,}g = \overline{1, G}$.

The equivalent optimized function, based on (2.22), has the form:

$$F\left(z_g\right) = \left[-\sum_{g=1}^{G} z_g + \sum_{j=1}^{G} \alpha_j \left(\sum_{g=1}^{G} c_{jg} z_g - 1\right)\right].$$

The transformation A for calculating the predictive estimate we construct on the basis of the variational procedure [12]:

$$\Delta_g^k F = F\left(z^k / z_g = 0\right) - F\left(z^k / z_g = 1\right), \tag{4.8}$$

where $\nabla^k F$—variation of the function F by the variable z_g on the kth iteration; $z^k = \left(\tilde{z}_1, \ldots, \tilde{z}_{g_1}, \ldots, \tilde{z}_G\right)\left(g_1 = \overline{1, G}, g_1 \neq g\right)$—a vector of randomized Boolean variables. Randomization allows in depending from the fulfillment of the condition

$$P\left(z_{g_1} = 1\right) = p_{z_{g1}} \le \tilde{\xi}, \quad g_1 = \overline{1, G}, \quad g_1 \neq g,$$

where $\tilde{\xi}$—a random variable uniformly distributed over the interval [0, 1], to get different sets of values of 0 and 1 for fixed values of z_g, and accordingly, different values of variation (4.8). For predictive evaluation, we will calculate 2 implementations (4.8):

$$\Delta_{1g} F\left(z_g\right) \text{ and } \Delta_{2g} F\left(z_g\right).$$

Additionally, we introduce random steps of movement in the predicted direction.

$$\tilde{b}_g = \begin{cases} 1, \text{ with distribution } P\left(\tilde{b}_g = 1\right) = p_{b_g}, \quad P\left(\tilde{b}_g = 0\right) = q_{b_g}, \\ 0, \text{ in otherwise} \end{cases}$$

$$\tilde{c}_g = \begin{cases} 1, \text{ with distribution } P\left(\tilde{c}_g = 1\right) = p_{c_g}, P\left(\tilde{c}_g = 0\right) = q_{c_g}. \\ 0, \text{ in otherwise} \end{cases}$$

At the first iteration, it is similar (4.6)

$$p^1_{b_g} = 0, 5, \, p^1_{c_g} = 0, 5.$$

At the following iterations, we have.

$$p^{k+1}_{b_g} = p^k_{b_g} + p^{k+1}_{c_g}\left[q^k_{b_g}\theta\left(\Delta^k_{1g}F \cdot \Delta^k_{2g}F\right) - p^k_{b_g}\theta\left(-\Delta^k_{1g}F \cdot \Delta^k_{2g}F\right)\right],$$

$$p^k_{c_g} = p^1_{c_g} = 0, 5,$$

where θ—characteristic function

$$\theta(a) = \begin{cases} 1, & \text{if} a > 0 \\ 0, & \text{if } a < 0. \end{cases}$$

The transformation A has the form

$$A = p^{k+1}_{b_g}\left[q^k_{z_g}\theta(\Delta^k_{1g}F \cdot \Delta^k_{2g}F) - p^k_{z_g}\theta(-\Delta^k_{1g}F)\right].$$

According to the probability value (4.7), the dominant solution option is determined (see clause 2.1). Thus, combining the 3 stages of optimization modeling, in accordance with the structural scheme (Fig. 4.3), we obtain an invariant description of the professional activity model that characterizes the choice of KIA by management of the production activities of an air company.

4.3 Inductive Logic Model of the Assessment the Causes of Deviations of Actual Values of Key Indicators of Activity from Planned Ones

The formation of an inductive logical model, as an invariant content of the basic tasks of controlling the maintenance of air transport, requires the following steps:

determining the set of evidence of the deviation of the actual values of KIA from the planned ones;

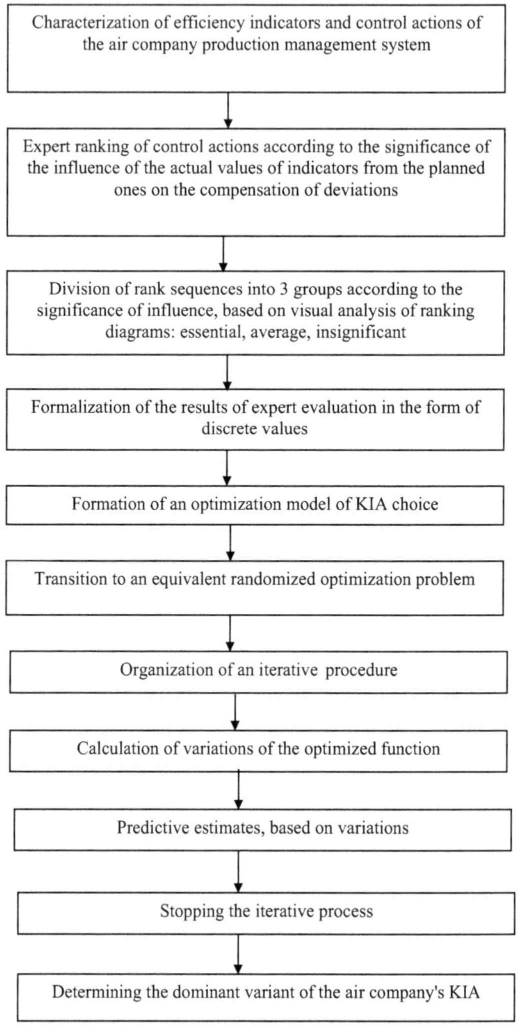

Fig. 4.3 Block diagram of the construction of an invariant description of the professional activity model that characterizes the choice of KIA for the management of the air company's production activities

determination of a set of hypotheses (no more than 3–4) about the reasons for the deviation, corresponding to each certificate about deviation of the actual values of KIA;

determination of a priori plausibility of hypotheses $P(1)$, $P(2)$,…of a posteriori plausibility, based on the obtained evidence α, β …, respectively, for each hypothesis, i.e., $P(1/\alpha)$, $P(2/\beta)$,…;

constructing a sequence of inductive conclusions, based on axioms, theorems and rules for calculating the probabilities of hypotheses [13] to identify the causes of deviations.

The first three stages are carried out on the basis of a study of the management system by KIA. To proceed to the main, fourth stage, it is necessary to use a number of theorems of inductive logic.

Let denote as $I(\alpha, \beta)$ the relation of the inductive conclusion between the statements α and β. The symbol itself reads as follows: The statement (hypothesis) α follows inductively from the statement (evidence) β. Instead of "the hypothesis α follows inductively from the evidence β", we can also say that the evidence β onfirms the hypothesis α, i.e., increases the value of its likelihood in comparison with the initial likelihood. Accordingly, the rule R_1 can be interpreted as follows: the statement α follows inductively from the statement β, $I(\alpha, \beta)$, if and only if the inequality $P(\alpha/\beta) > P(\alpha)$ is deductively deducible.

Literally, inequality $P(\alpha/\beta) > P(\alpha)$ means that the evidence β confirms the hypothesis α, since in its presence, the degree of likelihood of the hypothesis increases compared to the initial level.

If there is an inequality $P(\alpha/\beta) < P(\alpha)$, then it is customary to say that the evidence β does not confirm the hypothesis α, since it decreases the value of its likelihood in comparison with the initial level.

If there is an equality $P(\alpha/\beta) = P(\alpha)$, then it means that the evidence β turned out to be inductively neutral for the hypothesis α, since it does not increase or decrease the degree of its likelihood of the relative a priori level.

Combining together the cases of non-confirmation and inductive neutrality of hypotheses, we get from the rule R_1 that the statement α does not follow inductively from the statement β, i.e., $\neg I(\alpha, \beta)$, if and only if $P(\alpha/\beta) \leq P(\alpha)$.

The conjunction of arbitrary statements α and β of inductive logic will be denoted as $(\alpha\beta)$.

For a probabilistic assessment of the reasons for the deviation of the actual KIA values from the planned ones on the basis of an inductive logical model, we use a number of conclusions that follow from these theorems [13, 14].

1. Deviation $I(\alpha, \beta)$ is possible in the case of statements α and β that do not contradict each other;
2. The Plausible Hypothesis of Inductive Inference About the Reason for the Deviation of *H*, Following from the Evidence of *E*, is Confirmed by This Evidence;
3. If the analysis of the reasons for the deviation reveals the hypothesis H_2, which is alternative to H_1, then the estimate of the degree of likelihood of evidence *E* with respect to H_1 is higher than the probability estimate with respect to the alternative hypothesis H_2

$$P(E/H_1) > P(E/H_2).$$

4. In the case, when analyzing the reasons for the deviation, the hypothesis *H* is confirmed by the evidence of E_1 and E_2, and the a priori probability estimate is higher for E_1:

$$P(E_1) > P(E_2),$$

then this condition is also preserved for the relation of conditional probability estimates:

$$P(E_1/H) > P(E_2/H)$$

and it is necessary to choose the evidence E_1 as the reason for the deviation.

5. If the analysis of the reasons for the deviation reveals alternative mutually exclusive reasons and accordingly, evidence, then the one, whose a priori likelihood is greater than the priori likelihood of each of the alternative reasons, is confirmed.

In addition, to construct a scheme for identifying the causes of deviations, it is necessary to accept the condition that the set of all formed hypotheses is complete, and one of the hypotheses under consideration is true, i.e., $\sum_{i=1}^{n} P(H_i) = 1, n$ is the total number of hypotheses. We assume that all priori plausibility $P(H_i)$ are known before obtaining evidence, i.e., the occurrence of a deviation. A priori plausibility can be refined after receiving the certificate and turn into a posteriori plausibility.

Let there is the evidence E of KIA deviation. Based on this evidence, n plausible hypotheses $P(H_i)$ are formed, where $i = \overline{1, n}$, $\sum_{i=1}^{n} P(H_i) = 1$. We believe that all $P(H_i)$ are known because the existing experience of managing an air company. After the appearance of the evidence, the a priori plausibility changes to a posteriori, i.e., we get values $P\left(H_i / E\right)$, some of which may change in relation to the a priori ones both in the direction of increase and decrease, and some part may not change within the framework of the received evidence. It is obvious that those hypotheses for which the result $P\left(H_i / E\right) < P(E)$ are obtained, i.e., the evidence E does not confirm a number of a priori plausibilities should be rejected. In turn, this will require a revision of other, now a posteriori plausibility.

To obtain a constructive result and build an inductive model, it is necessary: (1) write out the entire series of a posteriori plausibility of the selected hypotheses; (2) rank this series and get the desired answer, i.e.,

$$P\left(H_i / E\right) < P(E) > P\left(H_{n-i} / E\right), \quad i = \overline{1, n},$$

where $P\left(H_i / E\right)$ is the maximum likelihood for the i-th hypothesis with respect to all other plausibility hypotheses $P\left(H_{n-i} / E\right)$.

A block diagram of the construction of an inductive logical model of the causes of deviations of the actual KIA from the planned ones is shown in Fig. 4.4.

Having received probabilistic estimates of the reasons for the deviation of the KIA, we will proceed to the construction of an invariant model of professional activity, associated with the optimization of management decision-making to compensate the deviations of the actual KIA values from the planned ones.

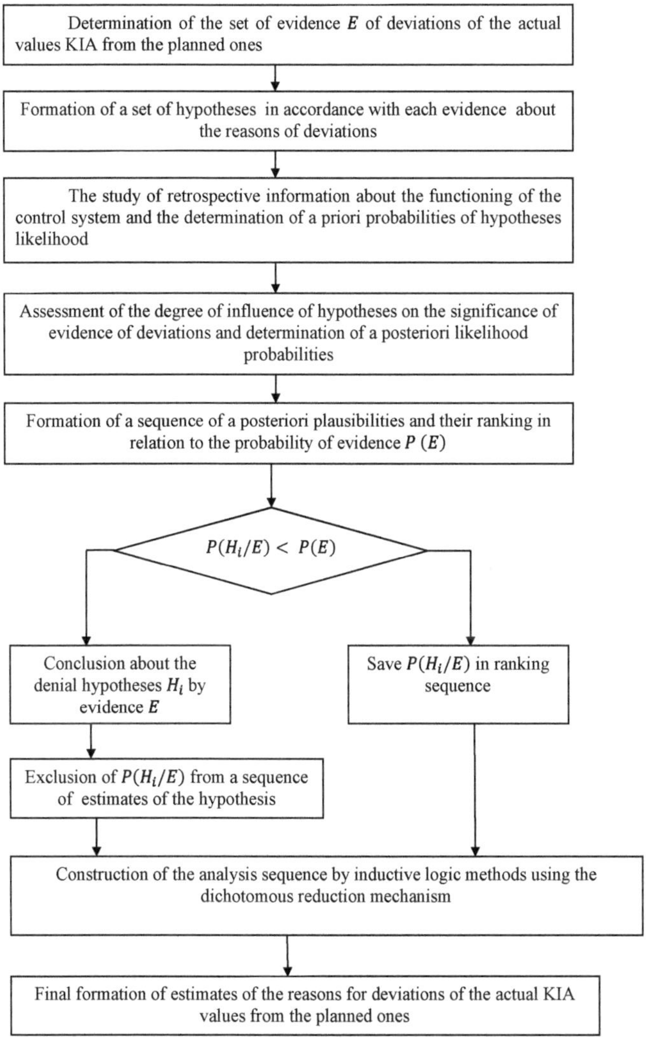

Fig. 4.4 Block diagram of the construction of an inductive logical model of the causes of deviations of the actual KIA values from the planned ones

4.4 Optimization of Acceptance of Management Decisions About Compensation of the Deviations of Actual Values of Key Indicators from Planned Ones

Using the model of probabilistic assessment of the causes of deviations, set out in clause 4.3 allows us to determine the a posteriori plausibility, based on the presence of one evidence E, that is, the distribution of conditional probabilities $P(H_i/E)$

of the influence on the deviation of the causes (hypotheses), $H_i, i = \overline{1, n}$, where $\sum_{i=1}^{n} P(H_i/E) = 1$. The evidence E determines the deviation from the planned values of some subset $G_1 \in G$ of the actual values KIA—$y_{g_1}, g_1 = \overline{1, G_1}$. Then, the H_ith hypothesis means the deviation of the indicator y_{g_1} and

$$P\left(y_{g_1}^{a} > y_{g_1}^{pl}/E\right) = P\left(H_{g_1}/E\right) = P(H_i/E),$$

where $y_{g_1}^{a}, y_{g_1}^{pl}$—the actual and planned values of the indicator y_{g_1}, respectively.

Based on the above meaningful statement, we will construct an invariant description of this model of professional activity in the form of an optimization problem (see Table 4.4).

We introduce alternative variables for choosing management decisions

$$z_j = \begin{cases} 1, \text{ if } j\text{th management action is used to compensate of deviations KIA;} \\ 0, \text{ in otherwise}, \quad j = \overline{1, J}. \end{cases} \tag{4.9}$$

From the results of the expert assessment of the significance of the influence of management actions on the change of indicators (4.1), we will select the values $c_j g_1, j = \overline{1, J}, g_1 = \overline{1, G_1}$. Then, the optimization criterion is the requirement to maximize the following objective function

$$\sum_{g_1=1}^{G_1} \sum_{j=1}^{J} P\left(H_{g_1}/E\right) c_{jg_1} x_j \to \max, \tag{4.10}$$

taking into account at the same time the significance of the influence of causes on the deviation of the indicators y_{g_1} and the significance of the influence of the j-th management action on the change of the indicator y_{g_1}.

In order to form the boundary requirements for the integral significance of the management actions on the indicator y_{g_1}, we will determine the first of the three groups, characterizing the essential significance in the form of the set J_{1g_1} from the partition condition according to the rank diagram (see clause 4.2.). In a formalized form, we obtain the following system of restrictions

$$\sum_{j=1}^{J} r_{jg_1} x_j \le \sum_{j \in J_{1g_1}} r_{jg_1}, g_1 = \overline{1, G_1}. \tag{4.11}$$

The optimization model combines the optimization criterion (4.10) and constraints (4.9), (4.11):

$$\sum_{g_1=1}^{G_1}\sum_{j=1}^{J} P\left(H_{g_1}/E\right)c_{jg_1}x_j \to \max,$$

$$\sum_{j=1}^{J} r_{jg_1}x_j \le \sum_{j\in J_{1g_1}} r_{jg_1}, \quad g_1 = \overline{1,\ G_1}, \tag{4.12}$$

$$x_j = \begin{cases} 1, \\ , \\ 0 \end{cases} \quad j = \overline{1,\ J}.$$

In order to use a randomized dichotomous reduction algorithm for solving the problem (4.12), similar to the one, given in clause (4.2), we define an optimized function equivalent to the problem (4.12):

$$F\left(z_g\right) = \sum_{g_1=1}^{G_1}\sum_{j=1}^{J}\sum_{j=1}^{J} P\left(H_{g_1}/E\right)c_{jg_1}x_j + \sum_{g_1=1}^{G_1}\alpha_{g_1}\left(\sum_{j\in J_{1g_1}} r_{jg_1} - \sum_{j=1}^{J} r_{jg_1}x_j\right).$$

As a result of the solution, we will find the set of management actions $x^*_{j_1}$, $j_1 = \overline{1,\ J_1} \in \overline{1,\ J}$, the most significant for management by deviations of the actual values of KIA $y^a_{g_1}$ from the planned ones $y^{pl}_{g_1}$, $g_1 = \overline{1,\ G_1}$. In the considered invariant description, the final stage of optimization modeling is the search for optimal values of the intensity of management actions $x^*_{j_1}$, $j_1 = \overline{1,\ J_1}$. To formulate the parametric problem of finding the extremum by probabilistic estimates, we determine the indicator $\hat{g}_1$, for which the value of $P\left(H_{\hat{g}_1}/E\right)$ is the maximum. Based on this indicator, we will form an optimization criterion

$$\left(y_{\hat{g}_1} = \varphi\left(x_{j_1}\right) - y^{pl}_{\hat{g}_1}\right)^2 \to \min. \tag{4.13}$$

Restrictions are set for other indicators and values x_{j_1}

$$\left(y_{g_1} = f_{g_1}\left(x_{j_1}\right) - y^{pl}_{g_1}\right)^2 \le \varepsilon_{g_1}, \quad g_1 = \overline{1,\ G_1}, \quad g_1 \ne \hat{g}_1,$$

$$x^{\min}_{j_1} \le x^{\max}_{j_1}, \quad j_1 = \overline{1,\ J_1}. \tag{4.14}$$

$$x^{\min}_{j_1} \le x_{j_1} \le x^{\max}_{j_1}, \quad j_1 = \overline{1,\ J_1}, \tag{4.15}$$

where ε_{g_1}—permissible deviations of indicators, established by expert method; $x_{j_1}^{\min}, x_{j_1}^{\max}$—minimum and maximum values of management actions due to the capabilities of the management system.

Combining the optimization criterion (4.13) and the constraints (4.14), (4.15) we have the following optimization problem:

$$\left(\varphi\left(x_{j_1}\right) - y_{\hat{g}_1}^{pl}\right)^2 \to \min, \tag{4.16}$$

$$\left(f_{g_1}\left(x_{j_1}\right) - y_{g_1}^{pl}\right)^2 \le \varepsilon_{g_1}, \quad g_1 = \overline{1,\ G_1}, g_1 = \hat{g}_1,$$

$$x_{j_1}^{\min} \le x_{j_1} \le x_{j_1}^{\max}, \quad j_1 = \overline{1,\ J_1}.$$

Let us form an equivalent optimization function for the problem (4.16)

$$F\left(x_{j_1}, \alpha_{g_1}\right) = -\left(\varphi\left(x_{j_1}\right) - y_{\hat{g}_1}^{pl}\right)^2 + \sum_{\substack{g_1=1 \\ g_1 \ne \hat{g}_1}}^{G_1} \alpha_{g_1}\left(\varepsilon_{g_1} - \left(f_{g_1}\left(x_{j_1}\right) - y_{g_1}^{pl}\right)^2\right).$$

For search of the optimal solution, we use gradient algorithm [2]

$$x_{j_1}^{k+1} = x_{j_1}^{k} + \gamma \left[\begin{array}{l} -2\left(\varphi\left(x_{j_1}^k\right) - y_{\hat{g}_1}^{pl}\right)\dfrac{\partial \varphi\left(x_{j_1}^k\right)}{\partial x_{j_1}} \\ + \displaystyle\sum_{\substack{g_1 = 1 \\ g_1 \ne \hat{g}_1}}^{G_1} 2\alpha_{g_1}^{k+1}\left(\varepsilon_{g_1} - \left(f_{g_1}\left(x_{j_1}^k\right) - y_{g_1}^{pl}\right)\right)\dfrac{\partial f_{g_1}\left(x_{j_1}^k\right)}{\partial x_{j_1}} \end{array}\right];$$

$$\alpha_{g_1}^{k+1} = \max\left\{0, \alpha^k - \lambda\left(\varepsilon_{g_1} - \left(f_{g_1}\left(x_{j_1}^k\right) - y_{g_1}^{pl}\right)^2\right)\right\}$$

Where

γ, λ—the values of the steps that determine the movement in the direction of the gradient;
$k = 1, 2, \ldots$—iteration numbers of the numerical procedure;
$x_{j_1}^1, \alpha_{g_1}^1, \gamma, \lambda$—the values of the optimized variables and quantities that are set as the initial conditions for searching for the optimal solution.

A block diagram of the construction of an invariant description of the model of professional activity for making management decisions is shown in Fig. 4.5.

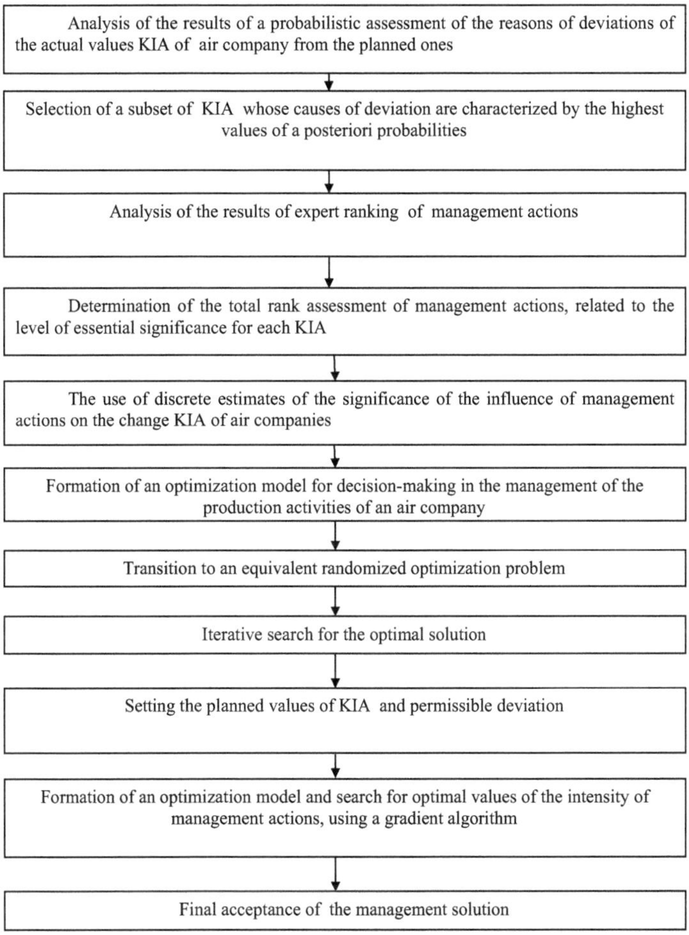

Fig. 4.5 Block diagram of the construction of an invariant description of a professional activity model for acceptance of the management solutions to compensate of deviations of the actual values KIA of the air company from the planned ones

References

1. Afanasyev VG (2009) Fundamentals of management activity. Part 2. Practical aspects of air company management, MSTU CA. Moscow, p 69
2. Gruzkov SA, Ostanin SY, Sugrobov AM et al (2005) Electrical equipment of aircraft: textbook for universities, vol 1. Moscow Power Engineering Institute, p 564
3. Lvovich IY, Lvovich YE, Frolov VN (2016) Information technologies of modeling and optimization: a brief theory and applications. Scientific Book, Voronezh, p 444
4. Yelisov LN, Baranov VV (1999) Management and certification in the aviation transport system. Air Transport, M
5. Gorlenko OA, Miroshnikov VV, Kukareko AN (2010) Formation of professional standards in the field of quality, based on competence models. Bull Bryansk State Tech Univ 3(27):91–98

6. Lapidus VA, Novosti M (2002) Universal quality (TQM) in Russian companies p 432
7. Ilyenkova SD (ed) (2009) Quality management: a textbook for university students, studying economics and management specialties, 3rd ed, reprint. Moscow, Unity-DANA, p 352
8. Stevens G (1989) Integrating the supply chain. Int J Phys Distrib Mater Manage 19(8):3–8
9. Teo TSH, King WR (1997) Integration between business planning and information systems planning: an evolutionary-contingency perspective. J Manage Inform 14(1):185–215
10. Ding H, Lim A, Rodrigues B, Zhu Y (2005) The over-constrained airport gate assignment problem. Comput Oper Res, p 1867–1880
11. Dmitriev ON, Examnew SH, Lubaeva SH (2007) Strategic management of aviation industry corporation of Russia. M.: KnoRus, p 566
12. Aidarkin DV (2012) Ensuring structural connectivity of modules of professional training of flight personnel in order to increase the level of flight safety: abstract of the dissertation of the Candidate of technical sciences. St. Petersburg, p 23
13. Svetlov VA (2006) Modern logic. St. Petersburg, Peter, p 400
14. Shuman AN (2004) Modern logic. Theory and practice. Minsk: Ekonompress, p 416

Chapter 5
Optimization of Personnel Training for Civil Aviation, Based on Models of Professional Activity in the Field of Development of Infrastructure Systems of Airports

The infrastructure of transport industries can be considered as an important component of the production infrastructure. The way, that it functions in a stable and efficient way, is a necessary condition for ensuring a noticeable and stable pace in the economic growth of the country, in ensuring the integrity of its territory, what is the spatial security and how the quality of life of people increases [1–6]. But at the same time, under the existing conditions, the characteristics and speed of modernization of transport industries, as well as the achieved technical and economic levels in the transport infrastructure in Russian Federation, do not allow us to meet the growing public needs for their development. There is a discrepancy between the needs and the insufficient level of how the transport infrastructure is developing, the efficiency and quality of its work, and the growing demand of the economy and people for transport services.

Models, demonstrating promising competencies in the development of airport infrastructure systems, should be focused on two main classes [7–11]:

- geographically distributed airport systems;
- internal facilities and complexes of the airport.

The most acceptable form of invariant description for the first class is situational models that allow us to assess the initial state of the airport infrastructure of the central hub and regional airport infrastructure and consider the possibilities of transition to a new state with better characteristics by making certain decisions on the development of these complex systems.

Among the internal facilities and infrastructure complexes of airports, the leading requirements are primarily imposed on radio communication systems, radio navigation and information systems [7, 8, 10, 12–15]. Therefore, it is necessary to distinguish professional activity models for the following components:

- radio communication systems as an integral part of an automated air traffic management system;
- intra-airport communication;
- radio navigation systems;

A. Borzova et al., *Modeling and Forecasting of Staffing in Civil Aviation*,
Springer Aerospace Technology, https://doi.org/10.1007/978-981-19-6238-7_5

- aviation electrical systems;
- airport information systems.

The final stage is the optimization transformation of problem-oriented models into components of educational programs.

5.1 Situational Simulation of Development of Geographically Distributed Infrastructure of the Airports

Let us consider two situational models [10, 16, 17]:

- airport infrastructure of the Moscow aviation hub;
- regional airport infrastructure.

The first model is based on the analysis of the state of the geographically distributed infrastructure of this class.

The characteristics, related to the objects of ground-based air transport infrastructure at airports, differ in a significant degree of wear. For the most part, this is due to artificial coverings on runways, taxiways and places, where there are aircraft parking lots and lighting equipment. Only about two-thirds are characterized by runways that have an artificial surface (the service life for asphalt concrete coatings of runways and related components at airfields is 10 years, for cement concrete—20–25 years). The wear of artificial coverings of airfields at some airports of the Moscow air hub is significantly more than 50%. Significant projects in the development of infrastructure components for airports in the Moscow air hub are the following (Fig. 5.1):

- Sheremetyevo airport—expansion and reconstruction of runways during the construction of taxiways with a high-speed descent;

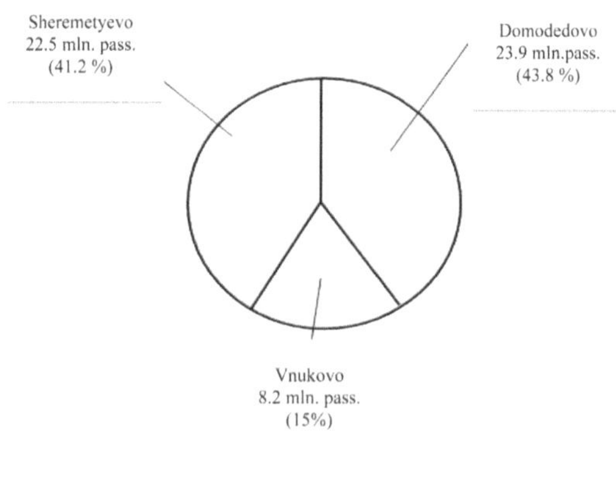

Fig. 5.1 Illustration of airport indicators of activity at the Moscow aviation hub

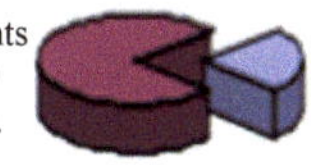

Fig. 5.2 Investments, related to the airports of the Moscow aviation hub (in general)

- Vnukovo airport—reconstruction of runways, taxiways and aprons of the airport, replacement of lighting equipment;
- Domodedovo airport—reconstruction of runways, taxiways and places, where aircraft parking is located.

Under the current conditions, there is a process of state financing in order to develop the infrastructure of airports, but at the same time it is necessary to attract additional capital investments in order to create a support network of hub airports. You can see in Fig. 5.2 that 171,805 million rubles were invested in MAH airports in 2006–2011 years. During 2006–2011 years, 3.3 rubles of additional funds were raised for 1 ruble of budget funds.

As a result of the fact that there is insufficient financing of the airport and airfield infrastructure for very long periods of time, the problems, associated with physical and moral wear, are not being solved. As one of the basic reasons for this situation, relating to the ground bases of civil aviation, it can be considered that insufficient amounts of investment are used, which are directed to support and develop airports. Based on the reconstruction of airfields, belonging to the major hub airports of Sheremetyevo, Domodedovo and Vnukovo, which carry out about 60% of air transportation volumes, it is possible to increase international transit through airports, increase their competitiveness within the aviation services markets and attract additional passenger traffic and extra-budgetary funds to expand terminal and refueling complexes. When the composition of the fleet of aircraft in Russian airlines changes and the intensity of their movement through MAH airports increases, due to insufficient investment in developing facilities in the ground infrastructure, it is inevitable that airports will not be able to ensure the balanced development of Russian Federations air transport systems [10, 11, 17, 18].

The degree of availability of modern equipment, related to aviation security, at Russian airports is insufficient under the existing conditions, while a significant level of wear can be observed (about 75%). Without urgent re-equipment in the material and technical base of airports in this area, there is an increase in the probability of terrorist threats in the field of civil aviation. At the same time, large additional investments are required to update the material and technical base, related to ensuring the safety of civil aviation flights.

Table 5.1 shows the current state at the airports, belonging to the Moscow aviation hub.

In order for there to be a phased development of facilities, related to the ground infrastructure of airports [19–23], in connection with the prospects for the development of transport systems in Russia, as well as a set of aircraft, the promotion of priority measures that include procedures for the reconstruction of runways and

Table 5.1 The current state at the airports, belonging to the Moscow aviation hub

Name of indicators	Characteristic		
	Vnukovo	Domodedovo	Sheremetyevo
Airfield infrastructure	2 depending intersecting runways: Runway-1—3000 × 60 Runway-2—3060 × 45	2 independing intersecting runways between the axes 2000 м: Runway-1—3500 × 60 Runway-2—3794 × 60	2 depending intersecting runways between the axes 283 м: Runway-1—3550 × 60 Runway-2—3700 × 60
The state of the airfield infrastructure	Runway-1 reconstruction is required	Runway-2 reconstruction is required	Runway-1 reconstruction is required
Throughput capacity of the runway, mln. pass in year	14.95	30.0	19.95
Throughput capacity of passenger terminal, mln. pass in year	15.0	21.0	33.0
The balance of passenger terminals to the airfield infrastructure	It is corresponding	It is not corresponding	It is corresponding with a reserve
Limitation of throughput capacity of airfield infrastructure	Business aviation and letter flights	It is absent	It is absent
Availability of car transport	Acceptable	Acceptable	An “backup” is required for the Leningrad highway
Availability of railway transport	It is secured	It is secured	It is secured
The approach of the airport to urban buildings and settlements	High density of building	It is absent	Insignificant

the replacement of lighting equipment, and also procedures for the modernization of existing capacities of facilities with effective transport services and meeting the demand of people and economic entities for air transportation and areas, equated to them, it is required, to ensure that:

- creation of support networks of airports, belonging to the federal significance, also applies to large transfer hubs and regional airports;
- carrying out the expansion, reconstruction and technical re-equipment of objects, related to ground infrastructure at airports, when carrying out transportation for new-generation aircraft of domestic production, when receiving foreign aircraft,

when introducing modern technologies, as well as new types of ground equipment when servicing passengers, shippers and consignees;

- attraction of extra-budgetary funds and budget funds in the constituent entities of the Russian Federation during the development of objects, related to the ground infrastructure of airports, this also applies to those, based on improving property management, which are in federal ownership.

The second situational model characterizes the state of the regional airport structure.

There is a complication of the situation on domestic (regional and local) air lines with problems that relate to the areas of transportation infrastructure: The number of airfields is currently 1.5 times less, if compared with 2000 years. At the end of the last decade, due to serious budget investments, the negative trend of compression of airport networks was stopped, but there has not yet been a revival of its lost part. Recently, in matters, related to the modernization of airport facilities, the state has been making about the main efforts to combine local airfields and landing sites into federal state-owned enterprises. As their main advantage, it can be noted that airport fees are artificially lowered (while restraining the growth of ticket prices), and also in the directions of financial flows to the ground infrastructure of local air lines, this makes it possible to take into account modern flight safety requirements. However, since various forms of ownership can be specified for objects in the ground infrastructure of air transport by region in the existing conditions, therefore, the pace of their consolidation is low.

Over the past 10 years, the average profitability of Russian airlines on domestic airlines has been negative, since their operational activities involved an extremely high financial burden. For example, during 2012 years, there was an increase in the cost of transportation in the domestic region of the country by 6.4%. Quite actively, there was an increase in the costs of renting and leasing aircraft (an increase of 27.7%), airport services (19%), aviation fuel and lubricants (9.7%). In 2012 years, the losses of Russian air companies on domestic air lines amounted to about 20 billion rubles (in the international transportation market, a profit of about 12 billion rubles was received) (Fig. 5.3).

As can be seen in Fig. 5.1, about 5.53 million passengers belong to regional and local transportation, which is 14.7% of the total number of passengers transported on domestic routes. As a result, among the key reasons, related to the negative development of the economy in regional transportation, in addition to the low effective demand of the population, we can note the unfavorable institutional conditions for their functioning. Also, an effective scheme for the interaction of trunk (network) and regional (feeder) carriers has not been created in Russian Federation. Together, this led to the decline in the fleet of regional and local aviation, which consisted mostly of Soviet-era aircraft (for Russian aircraft in regional aviation fleets, the share is about 67% (196 aircraft), among them only 12 belong to the new generation. They are represented by 10 An-148 jet and 2 An-140 turboprop. It should be noted that the situation with what prospect for the production of An-140 for commercial operation

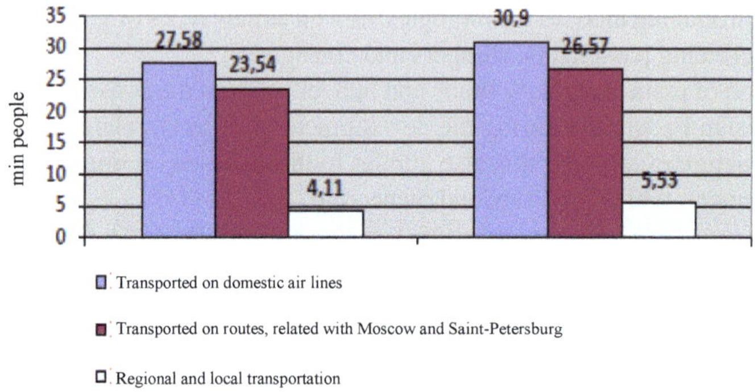

Fig. 5.3 Illustration of the share of regional traffic for domestic air lines

is currently not clear. The air companies did not have the funds to carry out updates, and there were no reliable business models that could attract external financing.

As the main reasons for the destruction of regional flight systems, you can specify:

- a significant decrease in the real incomes of the population and the level of effective demand for transportation in 1991–2000 years;
- absence in 1991–2000 years investments in the modernization and development of airport infrastructure;
- the existence of competition on the part of subsidized railway transport (in the existing conditions, 30% of the cost of transportation for common cars in long-distance trains is compensated from the federal budget of the Russian Federation);
- the existence of a shortage of a fleet of modern regional aircraft and aircraft for local lines, the absence of mass-produced jet and propeller aircraft with a capacity of up to 50 seats in the product line of domestic aircraft manufacturers.

The decrease in the number of local and regional airlines has led to the fact that until recently the single economic space in the inner region of the Russian Federation and its constituent federal districts was destroyed. The consequences of such a process are:

- the process of destroying the systems of intra- and interregional economic relations;
- the process of slowing down in the growth rates of the gross regional product of the subjects of the federation;
- the presence of a difficult financial situation of regional air companies and airports that do not have sufficient volumes of work in conditions of low effective demand for transportation.

During 2012–2013 years, The Government of the Russian Federation has taken unprecedented measures, related to the revival and promotion of the development of regional aviation. Since January 1, 2012, the program for subsidizing aircraft leasing has been launched. The amount of subsidies is from 15 to 30% of the cost of aircraft, depending on the type, year of manufacture, and characteristics. About 2 billion rubles were allocated for the implementation of the program last year, 2.15

billion rubles are planned to be allocated this year and 2.4 billion rubles are planned in the future. As part of the programs, applications have already been received for the purchase of 50 aircraft, and the total amount of subsidies is 1.2 billion rubles.

It is important to understand that in certain countries, regional and local air company support programs have been working productively for a long time. For example, in the United States, regional air transportation has been subsidized since 1978 at the expense of the federal budget within the Essential Air Service programs. Through the Department of Transport, subsidies are distributed for air companies, operating flights to small settlements that are located throughout the country. At the same time, there is subsidizing of commercially unattractive airlines that would not be able to receive regular flights without state support. Such a program, within the framework of which air services are provided for 153 US localities, operates in 32 states of the main territory of the country, as well as in Hawaii and Alaska. In the 2012 fiscal year, $143 million was allocated for it (in 2011 year—$54 million).

For the countries of the European Union, individually and within the single economic space of the union, we can note a rich history, associated with state regulation of air transportation. It is of considerable interest to consider the experience of state regulation of domestic air transportation, based on the fact that when making a single decision on state regulation within the European Union, preliminary work is carried out on common principles and approaches to explore and coordinate the opinions of a large number of participants in this process, taking into account the best practices and interests of all EU member states.

In the Russian Federation, it is necessary to develop and systematically develop a unified route network for regional and local air transportation, which could unite all the largest cities of the region on the basis of regular air traffic [16, 24–26]. Only the first steps have been taken toward the revival of regional aviation and local air traffic. But such sectors in air transport require not only that there is a "cash pumping", but also a comprehensive approach to solving many problems, in particular, the problems of producing modern aircraft in regional and small aviation.

Under the current conditions, 75% of all passengers, transported by airlines in the internal regions of the country, arrive in Moscow or make a departure from it, 61.5% of passenger turnover in transportation over distances up to 1000 km refers to air lines, and the point of departure or arrival in which is Moscow. When transporting on such airlines, 75% use medium-haul narrow-body aircraft. These lines are characterized by the fact that there is a stable all-season passenger traffic and no state subsidies are required.

The current route network on domestic airlines shows the formed model of socio-economic development of the Russian Federation with Moscow as the main economic center.

At the same time, many economists point out that the subsequent development of Russia requires an increase in the role of regions in the economy and the creation of new centers of active growth. In order to solve this problem, it is necessary to increase the mobility of the population and form stable interregional transport links. The formed model of transportation has a deterrent to the development of not only civil aviation, but also the economy of a large number of regions.

5.2 Modeling and Transformation of Professional Activity, Related to the Maintenance of Aircraft Radio Communication and Radio Navigation Systems, in Component of Educational Programs

The efficiency of air transport maintenance is inextricably linked with the problems of managing the training of relevant personnel, including those, who will work with systems of maintenance of radio communication and radio navigation [3, 6, 27–30].

In order to effectively train specialists in the modern system of higher education [31–41], it is necessary to develop a model of psychological and pedagogical competence [42–44], based on which it would be possible to achieve a comprehensive solution to the task [45–47] (Fig. 5.4).

Let us consider a number of competencies that specialists should have in aviation radio communications and navigation [7, 8, 48, 49].

General competencies:

- the ability to conduct the construction of oral and written speech in a logically correct way, in a reasoned and clear way; possession of the skills of rhetoric, conducting disputes, discussions and polemics;
- the ability to acquire new knowledge and use various forms of training, information and educational technologies;

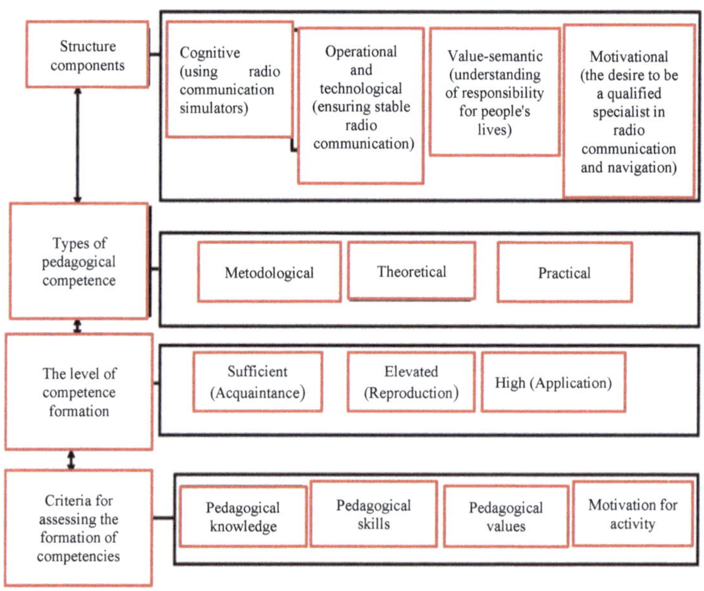

Fig. 5.4 Structural and functional model of competence

- the ability to apply knowledge on basic physical theories in solving emerging physical problems, to independently acquire physical knowledge, to understand the principles of operation of devices and devices, and this also applies to those that go beyond the competence of specific areas [45, 50–52].

Professional competencies:

- the ability to use knowledge of the laws and models of mechanics, vibrations and waves, electricity and magnetism, quantum physics, statistical physics and thermodynamics in order to solve professional problems [31, 53, 54];
- the ability to maintenance aircraft, engines and onboard systems, including radio and lighting equipment, based on the requirements of regulatory and technical documents;
- the ability to maintenance automated air traffic service systems, radio-electronic surveillance, navigation and communication systems, navigation and meteorological support for air traffic;
- readiness to perform work, related to information services for the maintenance of aircraft and aviation infrastructure facilities, air traffic management, air navigation services for flights and the use of airspace, based on computer equipment.

The process of forming competencies in the training of professional personnel in the area of radio communications and radio navigation can be provided on the basis of a set of lectures, practical exercises and relevant educational elements, included in the study module. The totality of the necessary educational resources provides a KAS-complex (knowledge, abilities, skills).

It is important to understand that competencies are developed mainly in an inconsistent way, but in an integral way, the formation of each competence occurs throughout the entire training cycle.

The task, associated with the distribution of educational resources in the educational process, can be presented on the basis of a model in which such designations are given:

Z_i, $i = 1 \dots I$—a set of competencies, related to level i;
$U_j(\tau), j = 1 \dots M$—a set of training tasks that are aimed at obtaining competencies (the development process), completed during the time τ;
$K_{i,s}$—s of sub-competencies, $s = 1 \dots S$, related to the competence of level i;
$U_{j,s}(\tau)$, $s = 1 \dots S$ is a set of training tasks, related to the implementation of s sub-competencies, $s = 1 \dots S$, related to the competence of level i, completed during the time τ.

In order to fulfill the curriculum, there is not enough knowledge and abilities that were acquired in the KAS-complex, and it is still necessary that some abilities, represented as a set of competencies $K_{i,s}$, should be shown. The totality of the competencies formed in this case, with proper interaction among the teaching staff, leads to the fact that there is a saving of time t, which is spent on creating these competencies for each of the students.

Let the KAS-complex be denoted as $Z = \{F_{v,b}\}$, where $v = 1 \ldots V$ is the set of training modules and b is the number of disciplines. The total volume for all KAS-complexes should not be more than the permissible time according to the curriculum.

In order to synthesize [55, 56] a work plan, it is necessary to follow such an algorithm. First, we have some educational resources (disciplines) that were provided for in the curriculum, and a set of KAS-complexes. There is an excess of the amount of resources, required to implement all KAS-complexes in the general form and scope of the curriculum. The formation of an optimal working curriculum is associated with the design of training periods that have an integral character, if we consider the development of competencies (those KAS-complexes that are implemented), and at the same time achieve ensuring their compliance with the planned volumes of educational resources.

Then, the hth complex ($h = 1 \ldots H$), associated with the formation of the sth competence, includes the following parameters: $Z(s, v, h) = \{D(s, v, h), T(s, v, h)\}$, where $D(s, v, h)$ is the coefficient of significance of the corresponding competence and $T(s, v, h)$ is the time of formation of the corresponding competence.

It is assumed that the coefficient of importance of competence takes values in the range from 0 to 1.

The coefficient of significance for the study of competencies is determined as follows:

$$Kr(v, i) = \frac{\sum_{s=1}^{S} \sum_{h=1}^{H} D(s, v, h)}{x_{1i} + x_{2i}}, \tag{5.1}$$

where x_{1i} and x_{2i} are set by experts: x_{1i} characterizes the degree of assimilation of educational materials and x_{2i} characterizes the importance of competence in terms of professional qualifications ($0 \leq x_{1i}, x_{2i} \leq 1$).

The coefficient of significance for studying professional competencies has the form [57–59]:

$$Krpr(v, i) = \frac{\sum_{s=1}^{S} \sum_{h=1}^{H} D(s, v, h)}{m_{1i} + m_{2i}}, \tag{5.2}$$

where m_{1i} and m_{2i} are set by experts: m_{1i} characterizes the degree of the use of competence in practice and m_{2i} characterizes the importance of competence for further development ($0 \leq m_{1i}, m_{2i} \leq 1$).

Optimization of the curriculum can occur in different ways [44, 60, 61]. In this book, we consider the task of synthesis in this way: to select the most important material for the curriculum from the point of view of professional activity and arrange it by semesters in an optimal way. The criterion that characterizes the importance of the educational material in the professional plan can be considered, for example, the following functionals:

- a criterion that maximizes the overall significance in the professional training of the modules, included in the curriculum:

$$\sum_{F_{v,b}\in Z} Kr(v,i) \to \max; \tag{5.3}$$

- the criterion on the basis of which the overall significance of the competencies [7, 8, 62–68], included in the curriculum, is maximized:

$$\sum_{F_{v,b}\in Z} Krpr(v,i) \to \max; \tag{5.4}$$

The criterion on the basis of which the optimal arrangement of modules for semesters will be determined, and we will consider the criterion by which the total value of time intervals between modules in the curriculum is minimized, which are related by information, taking into account the closeness of communication. It can be defined, based on the following dependencies:

$$\sum_{i=1}^{I}\sum_{v=1}^{V}\sum_{j=1}^{J}\sum_{h=1}^{H}\sum_{s=1}^{S} T(s,v,h)\cdot U_j(\tau)\beta_{v,v-1} \to \min, \tag{5.5}$$

where $\beta_{v,v-1}$ characterizes the closeness of the connection between the previous and subsequent module. It is determined by experts.

We will set four parameters: the total amount of training hours X, the total training time T, the result of training—R, levels of acquired competencies. The learning trajectory Z_i, T is considered as the desired variable. In general, the task, associated with the optimal management of the processes of competence formation, can be formulated as follows: For a given total amount of work U, which can be performed by the trainee, and the time, allotted for training T, it is necessary to determine the trajectory that maximizes the learning result Z. The task as a result can be written as follows:

$$\begin{aligned}
Kr(v,i) &= \frac{\sum_{s=1}^{S}\sum_{h=1}^{H} D(s,v,h)}{x_{1i}+x_{2i}},\ Krpr(v,i) \\
&= \frac{\sum_{s=1}^{S}\sum_{h=1}^{H} D(s,v,h)}{m_{1i}+m_{2i}} \sum_{F_{v,b}\in Z} Kr(v,i) \\
&\to \max,\ \sum_{F_{v,b}\in Z} Krpr(v,i),\ \sum_{i=1}^{I}\sum_{v=1}^{V}\sum_{j=1}^{J}\sum_{h=1}^{H}\sum_{s=1}^{S} T(s,v,h)\cdot U_j(\tau)\beta_{v,v-1} \\
&\to \min,\ \sum_{j=1}^{J}\sum_{s=1}^{S} U_{j,s}(\tau) \le U,\ \tau \le T,\ X \le T.
\end{aligned} \tag{5.6}$$

It can be considered as a problem of optimal management of educational processes, when it is necessary to distribute them over time at a fixed time and for a given amount of educational resources in such a way that the level of training is maximized and the required competencies are acquired.

Aviation telecommunications [7, 48, 49, 69] of civil aviation are considered as an integral part of automated systems, related to air traffic management (ATM). It provides interaction among air traffic management bodies and is used in the production, technological and commercial activities of air enterprises and their services. On the basis of aviation telecommunications, important production tasks are solved:

- the process of transmission by ATM centers to aircraft crews of relevant instructions, orders and messages, related to ensuring the safety and regularity of air traffic and receiving reports and messages from them at all stages of flights;
- the process of interaction of ATM centers, when there is ATM, flight planning and organization;
- conducting operational interaction of air company services;
- carrying out the transfer of administrative, management and production information;
- data transmission processes among various information systems that are used in the automation of production processes of civil aviation.

For aviation telecommunications [14, 48, 70–73], the following requirements must be met:

- timely establishment of communication;
- high level of reliability and uninterrupted communication;
- support of the required speed, when transmitting information;
- support of the required information protection during transmission;
- ensuring the economic efficiency of the functioning of telecommunications.

There are 3 types of RF civil aviation telecommunications: fixed (ground), mobile (air) and radio broadcasting.

The organization of fixed-line aviation telecommunications is carried out to provide the following functions:

- interaction among ATM centers;
- interaction among the services of CA air companies in the processes of conducting production activities;
- work of production and dispatching services and administrative and management personnel of the CA;
- transmission of meteorological and flight information;
- international flights of CA aircraft;
- interaction with Air Force authorities;
- data transmission.

Aviation mobile telecommunications [74, 75] can be arranged to provide the following functions:

- the immediate conduct of the controllers in the centers (points) ATM communications with the crews of aircrafts and data transmission throughout the flight and began taxiing before happened landing and ended taxiing;
- maintaining ATM radiotelephone and radiotelegraph communication with the crew of the aircraft that are in flight, by centers (points) ATM radiotelephone and radiotelegraph communication, it can occur and with the help of radio operators;
- maintaining by ATM centers (points), emergency rescue services of communication with the crews of aircraft that are suffering or have suffered a disaster.

Aviation radio can be arranged in order to provide the following functions:

- informing the crew that are in flight, for operational flight information service (OFIS);
- automatic transfer of information in the air station areas (ATIS);
- automatic transmission of meteorological information for aircraft crews, which are on the routes.

The technological radio communication of the airport is used for interaction during the production activities of the ATM bodies, for all airport services and air companies. The areas of application of technological radio communication of airports are:

- providing operational management of the functioning of ATM bodies, airport services and air companies during the planning, preparation and maintenance of aircraft flights;
- notifying the emergency rescue teams, when aviation accidents and incidents occur;
- obtaining the required information from enterprises and passengers, who use air transport.

The following can be noted as the main characteristics of the efficiency of communication facilities:

- the communication range (it depends on the parameters of the transceivers and antennas, the surrounding electromagnetic environment, weather conditions and terrain);
- the degree of support for the confidentiality of negotiations (interception and decryption of information is excluded);
- noise immunity characteristics;
- weight and dimensions;
- support for telephone network connectivity;
- support of various service functions;
- ensuring the reliability of maintenance in extreme conditions.

A characteristic feature of modern radio communication means is that they include a microprocessor that provides a set of service functions that expand the operational and tactical capabilities of radio stations:

- formation of groups of subscribers for independent operation on the same frequency;

- conducting scanning through several working channels;
- making a selective call to the subscriber;
- introduction of bans on transmission on busy channels;
- disabling and restoring stolen radio stations due to the fact that special radio commands are transmitted;
- ensuring the transfer of the formed program from one radio station to another in the field conditions, etc.

Radio stations, that are in the service of law enforcement agencies, are used for the most part. They are required in order to work in networks of both single-frequency and two-frequency simplexes. In the existing conditions, the used radio-technical means of navigation are complicated complexes of technical means that are designed to receive, process and transform information about the spatial and temporal position of moving objects. The radio navigation system is considered as one of the components of radio-technical navigation means.

In order to ensure aircraft driving along air routes, access to the areas of airfields and uncategorized landing approaches, angle-rangefinder systems of short-range navigation are used. Angle-measuring navigation devices (beacons, direction finders, radio compasses) are also used for the processes of navigation of moving objects. Among such means, we can note, for example, a radio navigation complex, which consists of a drive radio station and an automatic radio compass. It is designed to ensure flights along the route, the withdrawal of aircraft to the airfield, the implementation of a pre-landing maneuver and the execution of an uncategorized approach.

Based on the needs for professional training of aviation radio communication specialists, it is possible to determine offers for the model of an aviation radio specialist, which includes a target setting for training a radio communication specialist, who has a certain amount of knowledge, abilities and skills, as well as psychological readiness to perform their official duties.

Within the framework of the target settings, the teacher determines the primary, intermediate and final goals in the area of professional training of future radio specialists. After the goals of professional and practical training are determined, the teacher develops a professionally oriented training technology that most adequately meets the existing requirements, related to the training of aviation radio engineers [10].

Two components can be distinguished in the model—technological and informational (Fig. 5.5).

The technological component is necessary in order to increase the consistency in knowledge, practical skills of students, selection, actions related to the structuring and presentation of educational materials of special disciplines and control of the levels of their assimilation. It is a project of the educational process within the framework of a special discipline and is necessary in order to check the success of the application of professionally oriented learning technologies relative to the specified criteria, to react quickly when choosing forms, methods and means on the basis of which exercises are held, to redistribute time for training, the educational and material-technical base of the university, to organize interaction among students and teachers [31, 76, 77].

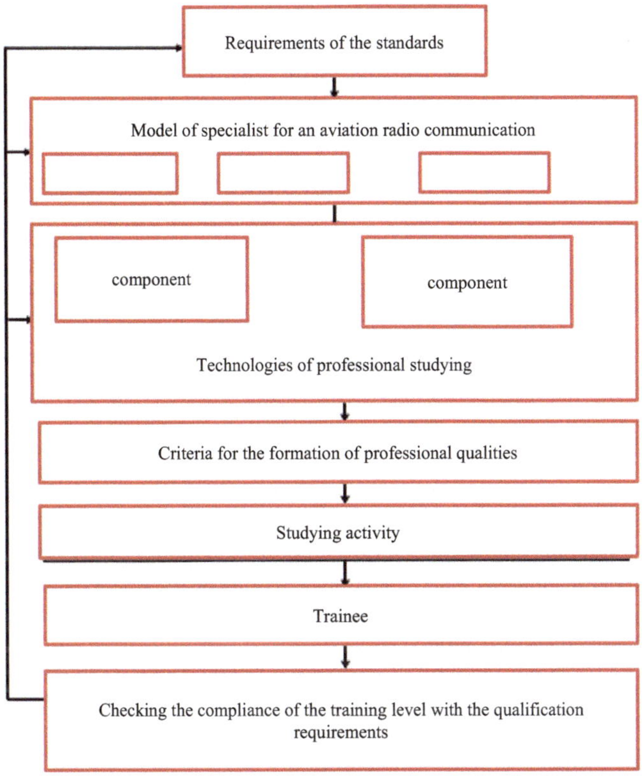

Fig. 5.5 Model of professional training of specialists for aviation radio communications

On the basis of the technological component, the features of the formation of professional and practical training of aviation radio communication engineers are determined, and it is assumed that there is a scientific justification for organizational approaches, related to determining the content of training, conducting training methods, managing cognitive functioning among students, as well as monitoring and evaluating learning outcomes. In the technological component, it is assumed that multi-level and individual training is being implemented.

The information component [78] is presented as a didactic complex, aimed at providing information for the study of special disciplines, and it includes:

1. computer textbooks, that cover the basics of training aviation radio specialists, the rules of conducting radio communications, on the basis of which knowledge can be acquired on the formation of aviation radio specialists, the basics of conducting radio communications;
2. a set of software products in order to form special and technical knowledge, abilities and skills. It provides opportunities to support multi-level and individual training.
3. training simulators;

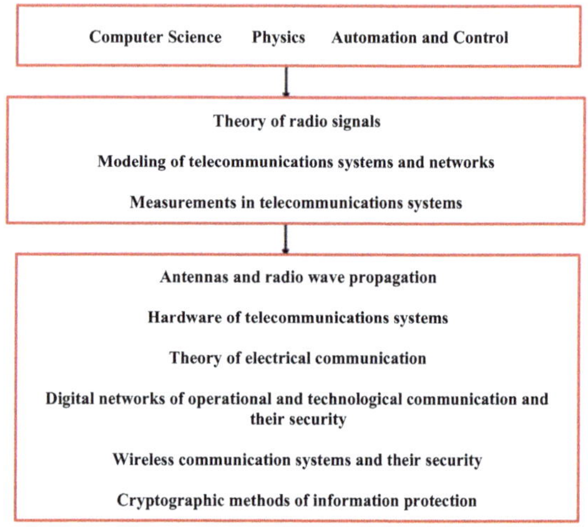

Fig. 5.6 Sequence of studying of disciplines in the training of an aviation radio engineer

4. means for monitoring relevant knowledge, abilities and skills, related to the transmission of information. This also applies to didactic materials that allow strengthening knowledge on conducting radio communications and radio codes, using printed and electronic media.

Figure 5.6 shows the sequence of studying disciplines in the training of an aviation radio engineer.

When forming learning technologies, the content of special disciplines is considered as distributed on the basis of the modular principle [79, 80].

After the first module has been studied, students acquire knowledge, related to the system of marking letters in the alphabet, when they are transmitted over the phone, using Morse code, they can acquire practical skills, related to the reception and transmission of alphabetic and digital radiograms at speeds of several tens of signs per minute. After the second module is studied, knowledge is acquired on the most frequently used service radio codes, the ability to transmit and receive information at speeds of several tens of signs per minute. After studying the third module, students acquire knowledge on the rules of radio communication for different modes, the procedure for maintaining accounting records, the official duties of the duty shift at radio complexes, practical skills in transmitting and receiving radio messages with alphabetic and digital content at higher speeds than in the second module and the ability to perform educational and professional tasks. When the fourth module is implemented, the solution of tasks, related to the connection of practical skills for the implementation of settings and work on radio means during the practical transmission of information among communication nodes at real distances, is carried out.

For example, due to the modular construction of the content of the discipline "Fundamentals of work at radio stations", it is possible to gradually develop for students a professional knowledge of the rules of radio communication, practical abilities and skills in transmitting and receiving information for telegraph and telephone modes, maintenance of radio equipment and performing the duties of specialists in radio departments.

Among the types of professional activities of an aviation radio engineer, the following can be noted:

1. design and construction of aviation radio communication facilities;
2. maintenance of aviation radio communication facilities;
3. bringing equipment into a state of readiness, conducting technical inspections and setting up radio-electronic equipment and devices. The specialization of an aviation engineer can be carried out only on the radio-electronic equipment of aircraft engineering.
4. management of the structural unit of aviation radio communications;
5. participation in research works, related to the development of new aviation radio communication systems.

Thus, we have 5 types of activities:

$$D_1 D_2, D_3, D_4, D_5. \tag{5.7}$$

For each type of activity, you can put it in compliance the labor functions:

$$\tau_{1m,}, \ldots, \tau_{im}, \ldots, \tau_{Im,} \tag{5.8}$$

where $i_m = \overline{1, I_m}$ is the numbering set of labor functions in mth type of activity, which are based on knowledge and abilities ($i = \overline{1, 5}$, $m = \overline{1, 5}$)

τ_{11}: control of the organization of accounting and storage of identification equipment, organization and accounting of the completion of improvements to aviation and radio-electronic equipment.

τ_{12}: implementation of technical management of design and survey work in the design of aviation radio equipment and author's supervision of its formation, commissioning and development of design capacities.

τ_{13}: to ensure a high level of feasibility of the designed objects of aviation radio communications, based on the use of the latest achievements of science and technology, most reasonable and economical design solutions.

τ_{14}: adoption of measures, aimed at improving the quality of design documentation and reduction of consumption of material resources in the construction of aircraft radio equipment, reducing the cost of their operation on the basis of improving the quality of design solutions.

τ_{15}: preparation of data for concluding contracts with customers for the development (transfer) of scientific and technical products, including justification of contractual prices.

τ_{16}: participation in the work of commissions on the selection of sites for the placement of aviation radio equipment, in the preparation of design assignments and in the organization of engineering surveys for the development of design estimates and other technical documentation.

τ_{21}: organization and control of work on metrological support for the maintenance and repair of aircraft equipment.

τ_{22}: organization of timely study and implementation of the requirements of the guidance documents on the maintenance of aviation and radio-electronic equipment.

τ_{23}: organization of practical exercises and internships for flight and technical personnel of the aviation and technical base, industrial practice of students and internships of teachers of educational institutions at the aviation and technical base.

τ_{24}: carrying out organizational, educational and methodological measures on the maintenance of aviation and radio-electronic equipment.

τ_{25}: implementation of engineering and aviation support for operational tests of aviation and radio-electronic equipment.

τ_{31}: analysis of the causes of failures and malfunctions of aviation and radio-electronic equipment, development of measures to prevent them.

τ_{32}: implementation of measures to reduce the duration of maintenance of aviation and radio-electronic equipment and reduce its cost, which does not worsen the quality of maintenance of this equipment.

τ_{33}: control of the technical condition and quality of service of aviation and radio-electronic equipment.

τ_{34}: not issuing a permit for the use of faulty and timely unverified diagnostic and ground maintenance tools for aviation and radio-electronic equipment that do not have accompanying documents, describing their technical condition.

τ_{35}: carrying out measures, aimed at economical expenditure of material and technical means and resources, allocated for the maintenance of aviation radio-electronic equipment.

τ_{41}: technical guidance and control over the work of specialists in aviation and radio-electronic equipment.

τ_{42}: taking measures to develop the production base of units servicing aviation and radio-electronic equipment, equipping them with advanced equipment, diagnostic and ground maintenance facilities, tools.

τ_{43}: organization and control of production activities of site of the processing and analysis of flight information.

τ_{44}: taking measures to ensure safe working conditions for specialists.

τ_{45}: control of the readiness of technical personnel to conduct practical exercises and internships with flight personnel on the study of aviation and radio-electronic equipment.

τ_{46}: removal from the performance of the production task of employees, who have shown unsatisfactory knowledge of aviation and radio-electronic equipment, the rules of its maintenance, violating or not fulfilling these rules.

τ_{47}: management of professional training of specialists in aviation and radio-electronic equipment, approval of work programs for the study of this equipment, verification of specialist's knowledge of the device and the rules of maintenance of aviation and radio-electronic equipment, and control of the admission of specialists to its maintenance.

τ_{48}: taking measures to equip technical classes on aviation and radio-electronic equipment with educational exhibits and literature.

τ_{49}: development and implementation of measures to train specialists of aviation and technical base units, servicing aviation and radio-electronic equipment, to master new equipment.

τ_{51}: conducting scientific research and development of aviation radio equipment on separate sections of topics as a responsible performer or together with a scientific supervisor, carrying out complex experiments and observations.

τ_{52}: collection, processing, analysis and generalization of scientific and technical information, advanced domestic and foreign experience, results of experiments and observations.

τ_{53}: participation in the preparation of plans and methodological programs of research and development, practical recommendations for the use of their results.

τ_{54}: preparation of reports on a topic or its section.

τ_{55}: participation in the implementation of research and development results.

For each type of activity, you can put it in compliance with the models of professional activity:

$$\mu_{1m}, \ldots, \mu_{jm}, \ldots, \mu_{Jm}, \tag{5.9}$$

where $j_m = \overline{1, J_m}$ is a numbering set of models, related to professional activity ($j = \overline{1, 5}, m = \overline{1, 3}$).

μ_{11}: a model of the activity of a specialist, associated with the design and construction of aviation radio communication facilities.

μ_{12}: a model of the personality of a specialist, associated with the design and construction of aviation radio communication facilities.

μ_{13}: a model of training a specialist, associated with the design and construction of aviation radio communication facilities.

μ_{21}: a model of the activity of a specialist in the maintenance of aviation radio communication facilities.

μ_{22}: a model of the personality of a specialist in the maintenance of aviation radio communication facilities.

μ_{23}: a model of training a specialist in the maintenance of aviation radio communication facilities.

μ_{31}: a model of the activity of a specialist, associated with the maintenance of aviation radio communication systems.

μ_{32}: a model of the personality of a specialist, associated with the maintenance of aviation radio communication systems.

μ_{33}: a model of training a specialist, associated with the maintenance of aviation radio communication systems.

μ_{41}: a model of the activity of a specialist in the management of a structural unit of aviation radio communications.

μ_{42}: a model of the personality of a specialist in the management of a structural unit of aviation radio communication.

μ_{43}: model of training a specialist in the management of a structural unit of aviation radio communications.

μ_{51}: a model of the activity of a specialist, associated with the scientific development of aviation radio communication systems.

μ_{52}: a model of the personality of a specialist, associated with the scientific development of aviation radio communication systems.

μ_{53}: a model of training a specialist, associated with the scientific development of aviation radio communication systems.

The competencies, specified in the educational standard, according to the types of activities are distributed as follows:

In the educational standard, the competencies

$$K_{1m}, \ldots, K_{rm}, \ldots, K_{Rm}, \tag{5.10}$$

where $r_m = \overline{1, R_m}$ is the numbering set of competencies [12, 79], required to perform the mth type of activity ($r = \overline{1, 5}$, $m = \overline{1, 8}$).

The specialist must:

K_{11}: receive and process information from various sources, using modern information technologies, analyze the received information, highlight the main thing in it and create new knowledge, based on it.

K_{12}: work with a computer, general and special-purpose software tools.

K_{13}: to analyze the state and prospects of development of both rocket and space technology as a whole as and its individual areas.

K_{14}: to know the basics of design, construction and production of orientation, navigation and stabilization devices for aircraft.

K_{15}: to know the basics of designing complex systems.

K_{21}: be able to apply reference materials.

K_{22}: to know the basics of the device of orientation, navigation and stabilization facilities of aircraft.

K_{23}: to know the basics of metrology, standardization and certification.

K_{31}: be able to use reference materials.

K_{32}: to know the basics of the device of orientation, navigation and stabilization facilities of aircraft.

K_{33}: to know the basics of metrology, standardization and certification.

K_{34}: know the technical regulations, interstate, national, industry standards and standards of the organization.

K_{41}: to work in a team.

K_{42}: to know the fundamentals of rocket and space technology.

K_{43}: to know the technical regulations, interstate, national, industry standards and standards of the organization.

K_{51}: to organize their work rationally, independently evaluate the results of their activities and possess the skills of independent work, including conducting research and development work.

K_{52}: to conduct a scientific search, independently or as part of a group, using special tools and methods for obtaining new knowledge.

K_{53}: to know the fundamentals of patent science.

K_{54}: to have the skills to create and edit professional texts.

K_{55}: to apply knowledge of the natural science and mathematical cycle, as well as practical experience in conducting scientific research.

K_{56}: to process and analyze the results of scientific and research works, find elements of novelty in the development.

K_{57}: to analyze the patent purity of the developed objects of professional activity.

K_{58}: to submit materials for registration of patents, prepare scientific articles for publication and prepare scientific and technical reports.

Relevant models

$$\mu_{1m}, \ldots, \mu_{sm}, \ldots, \mu_{Sm}, \tag{5.11}$$

where $s_m = \overline{1, S_m}$ is a numbered set of models of professional activity.

The multi-alternative choice consists in determining such a set of models from the sets (5.9), (5.11), which by the number of elements was minimal that corresponds to a limited educational resource, but at the same time each labor function (5.8) and competence (5.10) for all types of activities (5.7) correspond to at least one element from this set.

As a result, we obtain reduced sets of models for the D_mth type of activity μ^*_{jm}, μ^*_{sm} and form a single set for all types of activities $m = \overline{1, M}$

$$\mu = \mu^*_{jm} \bigcup_{m=1}^{M} \mu^*_{sm} \tag{5.12}$$

Thus, the reduced set includes the following models:

μ_{13}: a model of training a specialist, associated with the design and construction of aviation radio communication facilities.

μ_{21}: a model of the activity of a specialist in the maintenance of aviation radio communication facilities.

μ_{22}: a model of the personality of a specialist in the maintenance of aviation radio communication facilities.

μ_{23}: a model for training a specialist in the maintenance of aviation radio communication facilities.

μ_{33}: a model for training a specialist, related to the maintenance of aviation radio communication systems.

μ_{43}: a model for training a specialist for management of a structural unit of aviation radio communications.

μ_{53}: a model of training a specialist, associated with the scientific development of aviation radio communication systems.

We introduce a new numbering of the elements of the set (5.12)

$$\mu_1, \ldots, \mu_n, \ldots, \mu_N \tag{5.13}$$

That is,

$$\mu_1, \ldots, \mu_3, \ldots, \mu_6$$

and we will proceed to further use of the model-oriented approach.

The principle of optimal transformation of professional activity models into components of educational programs is based on an optimization model that connects many models of professional activity (5.13) with many components of educational programs

$$O_1, \ldots, O_v, \ldots, O_V, \tag{5.14}$$

where $v = \overline{1, V}$ is the numbering set of components of educational programs.

Educational programs:

$\boldsymbol{O_1}$: engineer for automated production management systems in the rocket and space industry.

$\boldsymbol{O_2}$: engineer for orientation, navigation and stabilization devices of aircrafts in the rocket and space industry.

$\boldsymbol{O_3}$: specialist in after-sales service of aviation equipment.

As a result, the models were distributed among educational resources in the following way:

$$\boldsymbol{O_1}: \boldsymbol{\mu_1}, \boldsymbol{\mu_2}, \boldsymbol{\mu_3}.$$

$$\boldsymbol{O_2}: \boldsymbol{\mu_2}, \boldsymbol{\mu_4}, \boldsymbol{\mu_5}.$$

$$\boldsymbol{O_3}: \boldsymbol{\mu_1}, \boldsymbol{\mu_4}, \boldsymbol{\mu_6}.$$

The transformation task is effective to show the elements of the set (5.13), when forming the elements of the set (5.14). To give the problem an optimization character, we introduce alternative variables:

$$x_{nv} = \begin{matrix} 0\;1\;1\;1\;1\;1 \\ 1\;1\;0\;0\;0\;1 \\ 0\;1\;1\;1\;1\;0 \end{matrix}, \quad n = \overline{1, 3}, v = \overline{1, 6}. \tag{5.15}$$

Next, using group expert evaluation, we will determine, based on the a priori ranking method, the rank sequence of models $\mu_n, n = \overline{1, N}$ for the formation of the component of the educational program $O_v - a_{nv}$, where a_{nv} takes integer values on the set $\overline{1, N}$ for the most significant model $a_{nv} = 1$.

As a result, we get educational resources in the form of the most appropriate distribution of a set of professional activity models:

$$O_1: \mu_1,$$

$$O_2: \mu_2, \mu_4,$$

$$O_3: \mu_1, \mu_4,$$

where

μ_1: a model for training a specialist, related to the design and construction of aviation radio communication facilities.

μ_2: a model of activity of a specialist in the maintenance of aviation radio communication facilities.

μ_4: a model for training a specialist in the maintenance of aviation radio communication facilities.

5.3 Modeling and Transformation of Professional Activities, Related to the Maintenance of Airport Information Systems, into Components of Educational Programs

One of the important components of the airport infrastructure is information systems [10, 81–86].

Let us consider a number of competencies that specialists in information systems should have [77, 87, 88]:

General competencies:

- the ability to present a modern picture of the world on the basis of an integral system of natural science and mathematical knowledge, to orient oneself in the values of being, life and culture;
- the ability to perceive, analyze, critically comprehend, systematize and synthesize information, obtained from various sources, predict, set goals and choose ways to achieve them;
- the ability and willingness to acquire new knowledge and use various forms of training, information and educational technologies;
- the ability to critically perceive information ("critical thinking"), analyze and synthesize it;
- possession of mathematical and natural science culture as a part of professional and universal culture;
- the ability to update the existing knowledge, abilities and skills, when making a decision and implementing it;
- knowledge of the basic concepts, principles, laws and rules of general and applied theory of systems.

Professional competencies:

- the ability to understand the essence and significance of information in the development of the modern information society, to be aware of the dangers and threats arising in this process, to comply with the basic requirements of information security, including the protection of state secrets;
- knowledge of the main methods, means and facilities of obtaining, storing and processing information;
- readiness to perform work, related to information services for the maintenance of aircraft and aviation infrastructure facilities, air traffic management, air navigation services for flights and the use of airspace, based on computer technology;
- the ability to use mathematical, analytical and numerical methods for solving professional problems, using ready-made software tools;
- the presence of computer skills as a means of information management;
- the ability and willingness to use information, obtained from global computer networks;
- ability and willingness to work with general-purpose software tools, when solving professional tasks;
- the ability to use programming languages and systems, computer modeling tools for solving various research and production tasks;
- readiness for continuous improvement of professional activities, decisions and developments in the direction of improving safety;
- the ability and readiness to organize and perform work on information support for the maintenance of aircraft and aviation infrastructure facilities, air traffic management, air navigation services for aircraft flights and the use of airspace, using computer technology.

Depending on the role of a person in the process of managing airport technological processes, computers, control and management tools, it is possible to divide all systems into two classes.

1. Information systems that ensure the collection and issuance of information in a convenient form in the course of technological or production processes. As a result of the corresponding calculations, it is determined which control actions need to be carried out so that the flow of the controlled process is the best [89–91]. The main role belongs to people, and the machine has an auxiliary role, by issuing the necessary information for them.
2. Control systems that allow providing, in addition to collecting information, issuing commands directly to performers or executive mechanisms, usually function for a real time scale, i.e., at the pace of technological or production operations. For control systems, an important role belongs to the computer, and a person controls and solves the most complex issues that, for some reason, cannot be solved by the computing tools of the systems.

The purpose of information systems [10, 16, 17] of automated airport management systems is to ensure that the operator receives information with high reliability and effective decision-making. As a characteristic feature in information systems, you can specify the operation of an electronic calculation machine in disconnected circuits of control. At the same time, there may be information systems of different levels.

Information systems should, on the one hand, provide reports on the normal course of processes at airports, on the other hand, information on situations that are caused by deviations from normal processes.

There are two types of information systems: information and reference (passive), they supply information to operators after their communication with the systems according to relevant requests, and information and advising (active), and they themselves periodically issue information to subscribers, intended for them.

Information and reference systems of electronic calculation machine are required only for the collection and processing of information about managed objects. Based on the information that is processed into a electronic calculation machine and provided in forms that are convenient for perception, the operator makes a decision about the method of managing of objects [24, 81].

Systems, related to data collection and processing, perform for the most part the same functions as in centralized control systems and are characterized by a higher level of their organization. The differences consist mainly in the qualitative nature.

In information and consulting systems, in addition to the fact that information is collected and processed, the following functions are performed:

- the rational technological mode of functioning for individual technological parameters of processes is determined;
- control actions are determined for all or individual process parameters;
- the values (values) of the settings in the local regulators are determined.

Data on technological modes and control actions are transmitted through the information display media in the form of recommendations for subscribers. Decision-making of subscribers is based on their own understanding of the process and experience. The schemes of the adviser system coincide with the schemes of the information collection and processing system.

The airport information system is a multi-level network of workstations, connected to each other in various ways (communication lines, using wireless technologies, the Internet).

According to these networks, all possible customs operations are carried out, accompanied by the exchange of large amounts of information. In most cases, the information contains confidential and secret data, such as personal data of customers, flight schedules, information about customers, staff, luggage and much more. In this regard, the urgency of protecting such a network from external influences and malicious attacks increases.

The control system [7, 24, 92] performs control functions for certain programs that provide in advance for actions to be taken in certain production situations. A person conducts general control and intervenes in such cases, when circumstances appear, that are not foreseen in the control algorithms. There are several varieties for control systems:

Supervisory control systems. The automatic control system, which operates in the supervisory control mode, is necessary in order to organize a multi-program mode of operation of an electronic calculation machine and is a two-level hierarchical system with great capabilities and it has increased reliability. The control program determines the evidence of the execution of programs and subroutines, and it manages the loading of devices of electronic calculation machine.

Systems of direct digital control. With the help of an electronic calculation machine, the optimal control actions are developed directly and control commands are transmitted to the actuators with the help of appropriate converters. The direct digital control mode makes it possible to apply more effective principles for regulation and management, to make a choice of their optimal variant; to implement optimizing functions and adapt to changes in the external environment and variable parameters of control objects; to reduce maintenance costs and unify the means of control and management.

Figure 5.7 shows a model of competencies of specialist for airport information systems.

In Fig. 5.8, the sequence of studying of disciplines in the training of specialists in airport information systems is given.

When forming learning technologies, the content of special disciplines is considered, distributed on the basis of the modular principle [69, 78, 93, 94].

The course "Computer science" can be characterized as occupying an important place for systems of multi-level training of specialists. Such a course will be the basis that ensures the introduction of information technologies into systems, related to vocational education.

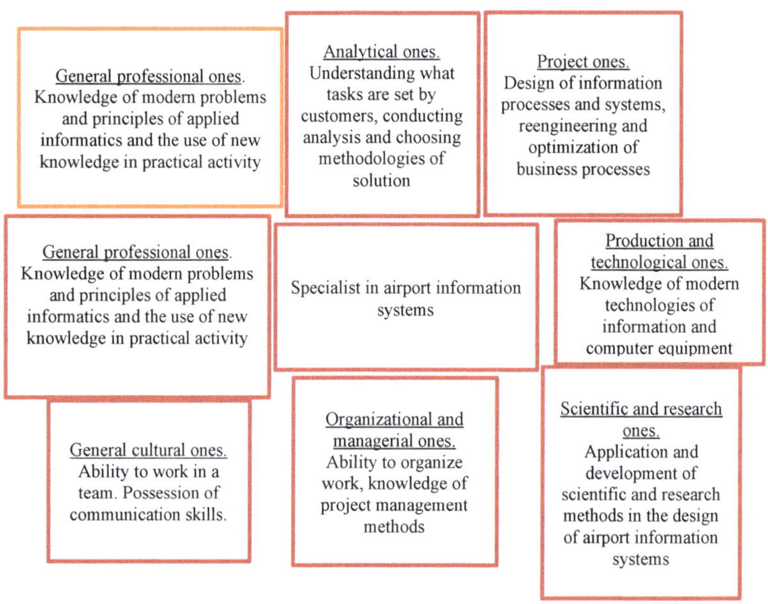

Fig. 5.7 Competence model of specialist in airport information systems

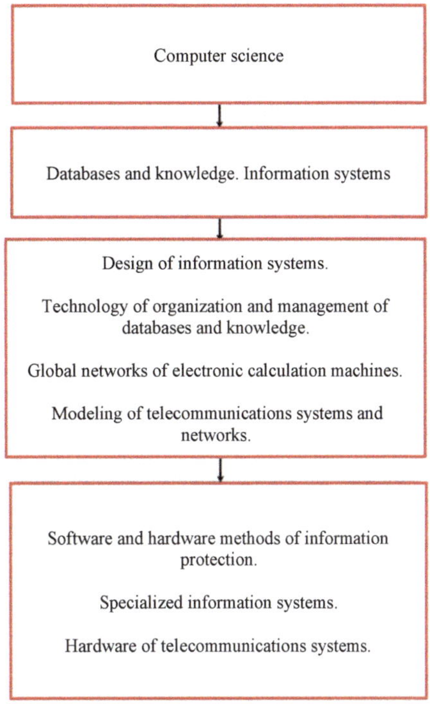

Fig. 5.8 Sequence of studying disciplines in the training of specialists in airport information systems

The purpose of the course "Designing of information systems" is that the main theoretical thesis of designing of automated information systems is studied; there is an introduction with what modern approaches to their development exist; the composition and content of technological design operations for various levels of hierarchy are studied; there is an introduction with existing tools in design, approaches to the formalized representation of design processes and methods of managing design processes.

Then, in the formation of knowledge, abilities and skills the discipline "Information systems" can be called as one of the leading roles. The main goal of the discipline is to obtain theoretical knowledge, related to the organization of automated information processing for organizations, to develop practical skills, related to the development of information systems (IS), that allow solving specific tasks (complexes) in the management of organizations and to get acquainted with the concepts of the development of automated information processing (AIP).

The discipline "Technology and organization of database management" is related to the basics of designing, managing and organizing a database, as the foundations of any information systems.

The discipline "Global networks of electronic calculation machines" is associated with the formation at students theoretical and practical skills, related to the organization and application of global networks for specific conditions.

In the process of studying this discipline, students should acquire knowledge on the basics of forming global networks.

Information protection is understood as the fact that in automated systems are created the organizational sets of tools, methods and measures that are designed to prevent of distortion, destruction or unauthorized teaching of protected information.

The purpose of teaching the discipline "Protection of information" is to form at students' theoretical and practical skills, related to the organization of protection of information for the conditions of specific applied areas, taking into account the cost aspect. In the process of studying the discipline, students should acquire knowledge, related to the basics of building systems of protection.

The discipline "Specialized information systems" is related to the study of airport information systems.

Among the types of professional activities of specialists, related to airport information systems, the following can be noted:

1. design and construction of airport information systems;
2. maintenance of airport information systems;
3. service of airport information systems;
4. management of the structural unit of the airport information system;
5. participation in scientific and research projects, related to the development of new airport information systems.

Thus, we have 5 types of activities:

$$D_1, D_2, D_3, D_4, D_5. \tag{5.16}$$

For each type of activity, you can match the labor functions:

$$\tau_{1m,}, \ldots, \tau_{im}, \ldots, \tau_{Im}, \tag{5.17}$$

where $i_m = \overline{1, I_m}$ is a numbering set of labor functions in the mth type of activity, which are based on knowledge and abilities ($i = \overline{1, 5}$, $m = \overline{1, 5}$).

τ_{11}: control of the organization of accounting and storage of information systems.

τ_{12}: conducting technical management of design and research work in the creation of airport information systems.

τ_{13}: ensuring a high technical and economic level of the projected components of airport information systems, based on the use of the latest achievements of science and technology, the most appropriate and economical design solutions.

τ_{14}: taking measures, aimed at improving the quality of design and estimate documentation and reducing the consumption of material resources in the formation of airport information system facilities, reducing the cost of their maintenance, based on improving the quality of design solutions.

τ_{15}: preparation of data for concluding contracts with customers for the development (transfer) of scientific and technical products, including justification of contractual prices.

τ_{16}: participation in the work of commissions on the selection of sites for the placement of elements of airport information systems, in the preparation of design tasks and in the organization of engineering surveys for the development of design estimates and other technical documentation.

τ_{21}: organization and control of work on metrological support for the maintenance and repair of airport information systems.

τ_{22}: organization of timely study and implementation of the requirements of the guidelines for the maintenance of airport information systems.

τ_{23}: organization of practical exercises and internships for specialists in airport information systems at the aviation technical base.

τ_{24}: conducting organizational, educational and methodological activities on the maintenance of airport information systems.

τ_{25}: implementation of engineering support for the ongoing maintenance tests of airport information systems.

τ_{31}: analysis of the causes of failures and malfunctions of elements of airport information systems, development of measures to prevent them.

τ_{32}: implementation of measures to reduce the duration of maintenance of airport information systems and reduce its cost.

τ_{33}: monitoring of the technical condition and quality of service of airport information systems equipment.

τ_{34}: not issuing a permit for the use of faulty and timely unverified diagnostic tools for airport information systems that do not have accompanying documents, describing their technical condition.

τ_{35}: carrying out activities that are associated with the economical expenditure of material and technical means and resources, allocated for the maintenance of airport information systems.

τ_{41}: technical guidance and control over the work of specialists in airport information systems.

τ_{42}: taking measures to develop the production base of departments, serving airport information systems, equipping them with advanced equipment, diagnostic means and tools.

τ_{43}: organization and control of production activities of the site of processing and analysis for flight information.

τ_{44}: taking measures to ensure safe working conditions for specialists.

τ_{45}: control of the readiness of the technical staff to conduct practical exercises and internships of specialists in the study of information systems.

τ_{46}: removal from the performance of a production task of employees, who have shown unsatisfactory knowledge of information systems, the rules for their maintenance and work with them, who violate or do not comply with these rules.

τ_{47}: management of professional training of specialists in airport information systems, approval of work programs for the study of such systems, verification of specialist's knowledge of the device and rules of maintenance of airport information systems, control and admission of specialists to their maintenance.

τ_{48}: taking measures to equip classes, related to the study of information systems with educational exhibits, literature.

τ_{49}: development and implementation of training activities, related to the maintenance of information systems.

τ_{51}: conducting scientific research and development of airport information systems on certain sections of topics as a responsible performer or together with a scientific supervisor, carrying out complex experiments and observations.

τ_{52}: collection, processing, analysis and generalization of scientific and technical information, advanced domestic and foreign experience, results of experiments and observations.

τ_{53}: participation in the preparation of plans and methodological programs of research and development, practical recommendations for the use of their results.

τ_{54}: preparation of reports on a topic or its section.

τ_{55}: participation in the implementation of research and development results.

For each type of activity, you can match the models of professional activity:

$$\mu_{1m}, \ldots, \mu_{jm}, \ldots, \mu_{Jm}, \tag{5.18}$$

where $j_m = \overline{1, J_m}$ is a numbering set of models, related to professional activity $(j = \overline{1, 5}, m = \overline{1, 3})$.

μ_{11}: a model of the activity of a specialist, associated with the design and construction of airport information systems.

μ_{12}: a model of the personality of a specialist, associated with the design and construction of airport information systems.

μ_{13}: a model for training a specialist, related to the design and construction of airport information systems.

μ_{21}: a model of the activity of a specialist in airport information systems.

μ_{22}: a personality model of specialist in an airport information systems.

μ_{23}: a model for training a specialist in the maintenance of airport information systems.

μ_{31}: a model of the activity of a specialist, associated with the maintenance of airport information systems.

μ_{32}: a model of the personality of a specialist, associated with the maintenance of airport information systems.

μ_{33}: a model for training a specialist, related to the maintenance of airport information systems.

μ_{41}: a model of the activity of a specialist for management of a structural unit of airport information systems.

μ_{42}: a personality model of a specialist for management of a structural unit of airport information systems.

μ_{43}: a model for training of a specialist for management of a structural unit of airport information systems.

μ_{51}: a model of the activity of a specialist, associated with the scientific development of airport information systems.

μ_{52}: a model of the personality of a specialist, associated with the scientific development of airport information systems.

μ_{53}: a model of training a specialist, associated with the scientific development of airport information systems.

The competencies, specified in the educational standard, are distributed, according to the types of activities, as follows:

In the educational standard, the competencies

$$K_{1m}, \ldots, K_{rm}, \ldots, K_{Rm}, \tag{5.19}$$

where $r_m = \overline{1, R_m}$ is the numbering set of competencies [12, 79, 95, 96], required to perform mth type of activity ($r = \overline{1, 5}$, $m = \overline{1, 8}$).

The specialist must [97, 98]:

K_{11}: to receive and process information from various sources, using modern information technologies, analyze the received information, highlight the main thing in it and create new knowledge, based on it.

K_{12}: to work with a computer, general and special-purpose software tools.

K_{13}: to analyze the state and prospects for the development of both rocket and space technology as a whole as and its individual areas.

K_{14}: to know the basics of design, construction and production of orientation, navigation and stabilization devices for aircraft.

K_{15}: to know the basics of designing complex systems.

K_{21}: be able to apply reference materials.

K_{22}: to know the basics of the arrangement of elements of information systems.

K_{23}: to know the basics of metrology, standardization and certification.

K_{31}: be able to apply reference materials.

K_{32}: to know the basics of the arrangement of the equipment of information systems.

K_{33}: to know the basics of metrology, standardization and certification.

K_{34}: to know the technical regulations, interstate, national, industry standards and standards of organization.

K_{41}: to work in a team.

K_{42}: to know the basics of specialized information systems.

K_{43}: to know technical regulations, interstate, national, industry standards and standards of organization.

K_{51}: rationally to organize your work, independently evaluate the results of your activities and possess the skills of independent work, including when conducting scientific research and experimental-design works.

K_{52}: to conduct a scientific search independently or as part of a group, using special tools and methods for obtaining new knowledge.

K_{53}: to know the basics of patent science.

K_{54}: to have the skills to create and edit professional texts.

K_{55}: to apply knowledge of the natural science and mathematical cycle, as well as practical experience in conducting scientific research.

K_{56}: to process and analyze the results of scientific and research works, find elements of novelty in the development.

K_{57}: to analyze the patent purity of the developed objects of professional activity.

K_{58}: to submit materials for registration of patents, prepare scientific articles for publication and prepare scientific and technical reports.

Relevant models

$$\mu_{1m}, \ldots, \mu_{sm}, \ldots, \mu_{Sm}, \tag{5.20}$$

where $s_m = \overline{1, S_m}$ is a numbered set of models of professional activity.

The multi-alternative choice consists in determining such a set of models from the sets (5.18), (5.20), which by the number of elements was minimal, which corresponds to a limited educational resource, but at the same time each labor function (5.17) and competence (5.19) for all types of activities (5.16) is corresponded to at least one element from this set.

As a result, we obtain reduced sets of models for the D_mth type of activity μ^*_{jm}, μ^*_{sm} and form a single set for all types of activities $m = \overline{1, M}$

$$\mu = \mu^*_{jm} \bigcup_{m=1}^{M} \mu^*_{sm} \tag{5.21}$$

Thus, the reduced set includes models:

μ_{13}: a model for training a specialist, related to the design and construction of aviation radio communication facilities.

μ_{21}: model of activity of a specialist in the maintenance of airport information systems.

μ_{22}: personality model of a specialist in the maintenance of airport information systems.

μ_{23}: model of training a specialist in the maintenance of airport information systems.

μ_{33}: model of training a specialist, related to the service of airport information systems.

μ_{43}: a model for training a specialist for management of a structural unit of airport information systems.

μ_{53}: a model of training a specialist, associated with the scientific development of airport information systems.

We introduce a new numbering of the elements of the set (5.21)

$$\mu_1, \ldots, \mu_n, \ldots, \mu_N \tag{5.22}$$

That is,

$$\mu_1, \ldots, \mu_3, \ldots, \mu_6$$

and we will proceed to further use of the model-oriented approach.

The principle of optimal transformation of professional activity models into components of educational programs is based on an optimization model that connects a variety models of professional activity (5.22) with a variety of components of educational programs

$$O_1, \ldots, O_\upsilon, \ldots, O_V, \quad (5.23)$$

where $\upsilon = \overline{1, V}$ is the numbering set of components of educational programs.

Educational programs:

$\boldsymbol{O_1}$: application and maintenance of automated systems for special purposes.

$\boldsymbol{O_2}$: infocommunication technologies and special communication systems.

$\boldsymbol{O_3}$: information security of telecommunications systems.

As a result, the models were distributed among educational resources in the following way:

$$\boldsymbol{O}_1\text{: } \boldsymbol{\mu}_2, \boldsymbol{\mu}_3, \boldsymbol{\mu}_4.$$

$$\boldsymbol{O}_2\text{: } \boldsymbol{\mu}_2, \boldsymbol{\mu}_3, \boldsymbol{\mu}_5.$$

$$\boldsymbol{O}_3\text{: } \boldsymbol{\mu}_1, \boldsymbol{\mu}_3, \boldsymbol{\mu}_6.$$

The transformation task is to show effectively the elements of the set (5.22), when forming the elements of the set (5.23). To give the problem an optimization character, we introduce alternative variables:

$$x_{n\upsilon} = \begin{matrix} 1\,1\,0\,0\,1\,1 \\ 1\,1\,0\,0\,0\,1 \\ 0\,0\,1\,1\,1\,0 \\ 1\,0\,1\,1\,1\,0 \end{matrix}, \quad n = \overline{1, 3}, \upsilon = \overline{1, 6}. \quad (5.24)$$

Then, based on the application of group expert evaluation, we will determine, based on the a priori ranking method, the rank sequence of models $\mu_n, n = \overline{1, N}$, when forming the component of the educational program $O_\upsilon - a_{n\upsilon}$. Here $a_{n\upsilon}$ can take integer values over the set $\overline{1, N}$, in the case of the most significant model $a_{n\upsilon} = 1$.

As a result, we get educational resources in the form of the most appropriate distribution of a set of models of professional activity:

$$\boldsymbol{O}_1\text{:}\boldsymbol{\mu}_2,$$

$$\boldsymbol{O}_2\text{: } \boldsymbol{\mu}_2, \boldsymbol{\mu}_5,$$

$$\boldsymbol{O}_3\text{: } \boldsymbol{\mu}_1, \boldsymbol{\mu}_3,$$

where

μ_1: a model for training a specialist, related to the design and construction of airport information systems.

μ_2: model of activity of specialist in the maintenance of an airport information systems.

μ_3: a model for training a specialist in the maintenance of an airport information systems.

μ_5: a model of the personality of a specialist, associated with the scientific development of airport information systems.

References

1. The State program (2014) Development of the transport system for 2016–2021. Approved by the Decree of the Government of the Russian Federation No. 303 of 15 April 2014
2. Elisov LN, Baranov VV (1999) Management and certification in the aviation transport system. Air transport, M
3. Levitin IE (2011) Current state and prospects of development of the Moscow aviation hub: report of the Minister of transport of the Russian Federation Igor Levitin March 28, 2011 at a meeting in Sheremetyevo airport, conducted by Vladimir Putin
4. Smirnov OV (2013) Ways of development of regional air transportation in the Russian Federation. Sci Transp Civ aviat 3(7):32–35
5. Smurov MYu (2013) About the main directions of the development of professional education in civil aviation in the conditions of innovative formation and modernization of the Russian economy. In: Formation and development of branch science and education in Russian air transport: Collection of abstracts of reports of the scientific conference. Federal State Unitary Enterprise State Scientific and Research Institute of Civil Aviation, Moscow, pp 19–23
6. Transport Strategy of the Russian Federation for the period up to 2030, approved by the Decree of the Government of the Russian Federation no. 1734-R
7. Borzova AS (2017) Improving the efficiency of management of radio communication and radio navigation maintenance systems in air transport on the basis of structural and functional modeling of the competence of professional activity. Econ Manage Control Syst 25(3.1):188–196
8. Borzova AS (2017) Training of specialists in aviation telecommunications. Bull Voronezh Inst High Technol 2(21):161–163
9. Borzova AS, Tsipenko VG (2017) Forecasting and optimization of the personnel training system in the field of maintenance of air transport, based on the principles of duality and expert-statistical modeling. Model Optim Inf Technol 2(17). https://moit.vivt.ru/wpcontent/uploads/2017/05/BorzovaZipenko_2_17_1.pdf. Data of access is 12 Feb 2018
10. Borzova AS (2017) Development of airport infrastructure on the basis of optimization modeling of the professional activity of an aviation radio engineer. Econ Manage Manage Syst 26(4):94–100
11. Borzova AS (2017) Expert-statistical formation of prognostic and evaluation models in the area of maintenance of air transport. In: New science as a result of innovative development of society: collection of articles of the International scientific and practical conference. Sterlitamak, pp 18–24
12. Bolelov EA (2016) Ensuring meteorological safety of aircraft flights due to the integration of methods of radiosonde tracking. Probl Saf Russ Soc (2):117–126
13. Vereshchaka AI, Olenyuk PV (1996) Aviation radio equipment: textbook. Transport, M., p 344
14. Gaskarov DV, Golinkevich TA, Mozgalevsky AV (1974) Forecasting the technical condition and reliability of radioelectronic equipment. Sov Radio, M., p 224

15. Gorshkov SA, Sedyshev SYu, Solonar AS, Yarmolik SN (2015) Fundamentals of radar: lecture notes. Chapter 3, VA RB, Minsk, p 180
16. Borzova AS, Zheleznaya IP (2013) Analysis of the state of the airport infrastructure of the Moscow aviation hub. Sci Bull Moscow State Tech Univ Civ Aviat (197):107–110
17. Borzova AS, Zheleznaya IP (2015) To the question about development of regional airport infrastructure. Sci Bull Moscow State Tech Univ Civ Aviat 217(7):23–26
18. Krasnov SI (2013) Main directions of improvement of training of aviation personnel in the Ulyanovsk higher aviation school of civil aviation (the Institute) and its affiliates. In: Formation and development of industrial science and education on the Russian air transport: collection of abstracts of reports of the scientific conference. Federal State Unitary Enterprise State Scientific and Research Institute of Civil Aviation, Moscow, pp 19–23
19. Eliseev BP (2016) Modern engineering education as the most important component of technological modernization of Russia. In: Innovative aspects of the development of civil aviation (Aviatrans-2016), pp 17–19
20. Spitsyn VV (2011) Features of innovative development of high technology and medium technology industries in Russia. Bull Tomsk State Univ (342):166–172
21. Chinyuchin YuM, Polyakova IF (2004) Fundamentals of technical maintenance and repair of aviation equipment: a textbook. Part I. MSTU CA, Moscow, p 81
22. Barnhart C, Belobaba P, Odoni AR (2003) Applications of operations research in the air transport industry. Transp Sci 368–391
23. Zachman J (1987) A framework for information systems architecture. IBM Syst J 276–292
24. Afanasyev VG (2009) Fundamentals of management activity. Part 2. Practical aspects of airline management. MSTU CA, Moscow, p 69
25. Borzova AS (2017) Calculation of the optimal labor intensity of the variable structure of the training course. In: Modeling and optimization in automated systems: materials of the All-Russian scientific school. Voronezh, Part 1, pp 31–37
26. Borzova AS, Zheleznaya IP (2012) Key indicators of activity of the air company. Sci Bull Moscow State Tech Univ Civ Aviat (181):35–38
27. Borzova AS (2017) About the opportunities for the development of aviation radio communications. In: Generation of the future-2017: the view of young scientists: a collection of scientific articles of the 6th Youth scientific conference. Kursk, pp 12–16
28. Lezhankin BV, Malisov NP (2009) Automated system for processing radar images of a meteorological navigation radar station of an aircraft. In: Proceedings of the 1st scientific and practical conference of teachers, researchers and postgraduates. Irkutsk, pp 165–172
29. Lushnikov AS, Tarasov SN (2012) Ground radio electronic means of ensuring aircraft flights: textbook. UI CA, Ulyanovsk, p 46
30. Perevezentsev LT (1999) The main directions of development of radar systems for air traffic management. Bull Nat Aviat Univ (1):114–119
31. Preobrazhensky AP, Choporov ON (2017) Management of the formation of competencies in the preparation of students. Mod Stud Soc Probl (Electron Sci J) 8(4–2):247–251
32. Preobrazhensky AP, Choporov ON (2017) Adaptation of students as young employees of the organization. Mod Stud Soc Probl (Electron Sci J) 8(6–2):161–164
33. Preobrazhensky AP, Choporov ON (2015) Analysis of the characteristics of the training of modern highly qualified engineering personnel. World Sci Discoveries 9.2(69):676–680
34. Preobrazhensky AP, Choporov ON (2017) The quality of education in higher education. In: Anthropocentric sciences: an innovative view of education and personal development: materials of the V international scientific and practical conference. Venice, Italy, pp 194–196
35. Preobrazhensky AP, Choporov ON (2017) The methodology of accounting for achievements, that allows increasing the motivation of students. Mod Stud Soc Probl (Electron Sci J) 8(6–2):165–168
36. Preobrazhensky AP, Choporov ON (2017) Optimization of the educational process, based on the use of information technologies. Mod Stud Soc Probl (Electron Sci J) 8(4–2):243–246
37. Preobrazhensky AP, Choporov ON (2015) Features of the use of information technologies in the training of modern specialists. World Sci Discoveries 9.2(69):670–675

38. Preobrazhensky AP, Choporov ON (2017) Application of statistical methods in enterprise management. Sci Krasnoyarsk Reg 6(1–2):273–278
39. Preobrazhensky AP, Komkov DV, Penchev GA, Vinyukov MS, Petrashchuk GI (2010) Problems of training specialists in modern high school. Mod Stud Soc Probl (1):66–67
40. Preobrazhensky AP, Choporov ON (2015) The role of student self-government as an innovative process. World Sci Discoveries 7.7(67):2664–2668
41. Preobrazhensky A.P. Management of innovative processes in educational organizations / A.P. Preobrazhensky, O.N. Choporov // Modern studies of social problems (electronic scientific journal), 2017, Vol. 8, no. 4–2, p. 252–256.
42. Zeer EF (2006) Psychology of professional development. Academy, M., p 240
43. Zernov VA, Lvovich YaE, Sorokin SO (2014) Optimization of the development of the non-state sector of higher education, based on the results of monitoring and rating assessment. Bull Voronezh State Univ Ser Probl High Educ (4):22–26
44. Zernov VA, Lvovich YaE, Mikhel AA (2014) Management of volume and resource characteristics of the higher education system, taking into account the potential of the university. In: Proceedings of the international lecture hall, dedicated to the 30th anniversary of the Department «Systems of automated design and information systems» of the Voronezh state technical university. Voronezh, pp 113–117
45. Zimnaya IA (2004) Key competencies as the effective and target basis of the competence approach in education. Research Center for Quality Problems of Training Specialists, M., p 38
46. Lobanova NN, Kosarev VV, Kryuchatov AP (1997) Professional competence of a teacher. St. Petersburg.: «SamVen», p 107
47. Lvovich YaE, Mikhel AA (2014) Structuring the management of resource and volume characteristics of the higher education system, using means of optimal choice. Bull Voronezh State Tech Univ 10(4):13–16
48. Evdokimov VD, Pavlov KM (1990) Designing of radio receiving devices. Communication, Moscow, p 12
49. Sadomovsky AS (2007) Receiving and transmitting radio devices and communication systems: a textbook. Ulyanovsk State Technical School, Ulyanovsk, p 243
50. Gorlenko OA, Miroshnikov VV, Kukareko AN (2011) Competence models of specialists in the field of quality. Competence 2(83):23–27
51. Gorlenko OA, Miroshnikov VV (2011) Coordination of bachelor's and master's competencies with the requirements of professional standards. Eng Educ (7):68–73
52. Gorlenko OA, Miroshnikov VV, Kukareko AN (2010) Formation of professional standards in the field of quality on the basis of competence models. Bull Bryansk State Tech Univ 3(27):91–98
53. Marienkin EV (2012) Management of aviation personnel on the basis of a competence model. In: Civil aviation: XXI century: materials of the 4th international conference. UI CA, Ulyanovsk, pp 132
54. Tatur YuG (2004) Competence in the structure of the model of quality for training specialists. High Educ Today (3):71–75
55. Sokolov NK (2012) Synthesis of optimal learning trajectories. In: Mechanical engineering and computer technologies (1), p 26
56. Tolmachev SA (1993) Presentation and synthesis of the optimal logical structure of educational material in the automated training systems. Cybern Syst Anal (1):183–186
57. Borzova AS (2017) The conceptual basis of the model-based approach to forecasting and optimization of the system of personnel training in the area of maintenance of air transport. Econ Manage Control Syst 24(2.1):188–194
58. Borzova AS, Tsipenko VG (2017) Multi-alternative optimization in the formation of educational resources, focused on models of professional activity. Model Optim Inf Technol 1(16). https://moit.vivt.ru/wp-content/uploads/2017/02/BorzovaZipenko%20_1_17_2.pdf. Data of access is 12 Feb 2018
59. Borzova AS, Tsipenko VG (2017) Multi-alternative choice of models of professional activity in the area of air transport maintenance. In: Fundamental and applied research: problems and

results: collection of materials of the XXXIII International scientific and practical conference. Novosibirsk, pp 46–50
60. Zhazha EYu, Nikolaev AB, Stroganov DV, Treshchetkina EYu, Prikhodko LV (2012) A formalized model of the curriculum in the task of optimizing of an individual educational trajectory. Sci Educ (11):351–361
61. Lapshina ML, Leshcheva NA, Lvovich YaE (2014) Optimization modeling of the structure of management accounting and budgeting of the main directions of school development. Econ Manage Control Syst 12(2):44–50
62. Borzova AS (2017) Optimal transformation of models of professional activity into components of educational programs. In: New technologies in education: materials of the XXV International scientific and practical conference. Moscow, pp 6–9
63. Borzova AS (2017) Optimization modeling of the choice of key indicators of activity in the tasks of air company management. In: Intelligent information systems: materials of the All-Russian scientific and practical conference with international participation. Voronezh, Part 2, pp 51–54
64. Borzova AS (2018) Optimization modeling of professional activity of specialists in airport information systems. Model Optim Inf Technol 6(1). https://moit.vivt.ru/wp-content/uploads/2018/01/Borzova_1_1_18.pdf. Date of access is 12 Feb 2018
65. Borzova AS (2017) Optimization of the components of the training content, based on expert analysis with a focus on the standards of a new generation. Model Optim Inf Technol 3(18). https://moit.vivt.ru/wp-content/uploads/2017/08/Borzova_3_1_17.pdf. Date of access is 12 Feb 2018
66. Borzova AS (2017) Features of aviation radio communication modeling. In: The generation of the future-2017: the view of young scientists: a collection of scientific articles of the 6th Youth scientific conference. Kursk, pp 17–20
67. Borzova AS (2017) Features of training specialists, related to the maintenance of aviation radio communication and radio navigation systems. In: Generation of the future-2017: the view of young scientists: a collection of scientific articles of the 6th Youth Scientific Conference. Kursk, pp 8–12
68. Borzova AS (2017) Planning of personnel training in the area of air transport maintenance, based on information technology of predictive modeling. In: Information technologies in education: materials of the IX All-Russian scientific and practical conference, pp 133–137
69. Aidarkin DV (2012) Ensuring structural connectivity of modules of professional training of flight personnel in order to increase the level of flight safety: abstract of the dissertation of the candidate of technical sciences. St. Petersburg, p 23
70. Vukolov EA (2008) Fundamentals of statistical analysis. In: Workshop on statistical methods and research of operations, using STATISTICA and EXCEL packages: a textbook. FORUM, Moscow, p 464
71. State standard R 54073-2010 (2011) Power supply systems for aircraft and helicopters. General requirements and standards of electricity quality. Standartinform, Moscow, p 33
72. Gruzkov SA, Ostanin SYu, Sugrobov AM et al (2005) Electrical equipment of aircraft: textbook for universities, vol 1. Moscow Power Engineering Institute, p 564
73. Davydov PS, Sosnovsky VA, Khaylovich IA (1984) Aviation radar: a reference book. Transport, M., p 233
74. Efimov AV (2004) Aviation radio electronics: textbook. In: Handbook for universities. UI CA, Ulyanovsk, p 219
75. Marienkin EV (2014) Methods of assessing the state of a complex system «aviation personnel» in the management process: abstract of the dissertation of the candidate of technical sciences. Moscow, p 19
76. Preobrazhensky YuP, Preobrazhenskaya NS, Lvovich IYa (2013) Media competence of a modern teacher. In: Secondary vocational education (12), pp 43–45
77. Preobrazhensky YuP, Preobrazhenskaya NS, Lvovich IYa (2013) Some aspects of informatization of educational institutions and the development of media competence of teachers and managers. Bull Voronezh State Tech Univ 9(5–2):134–136

78. Lavrentiev GV (2002) Designing the content of the operational module as part of the educational complex. Bull Bryansk State Univ (2):63–65
79. Bondarev YaP, Lvovich YaE (2013) Intellectualization of management of change in the activities of the university on the basis of monitoring-rating information. Mod Probl Sci Educ (3):13
80. Bondarev YaP, Lvovich YaE (2013) Management of changes in the values of ratings subjects and objects of the educational system, based on the process of intellectualization. Mod Probl Sci Educ (2):226
81. Amolin KC, Antal EI, Vasiliev VI, Granchina GO (2009) Control scheme of an autonomous group of unmanned aerial vehicles. Stoch Optim Comput Sci 5(1–1):157–166
82. Karpov EG, Koshkarev AC, Tikunov VC et al (2005) Geoinformatics. Publishing Center Academy, M., p 480
83. Kucheryavye AA (2004) On-board information systems: course lecture. In: Mishin VA, Klyuev GI (eds). Ulyanovsk State Technical School, Ulyanovsk, p 504
84. Lvovich IYa, Lvovich YaE, Frolov VN (2016) Information technologies of modeling and optimization: a brief theory and applications. Science Book, Voronezh, p 444
85. Pleshakova OA (2004) Mathematical models of organizations of production activities of a typical air company. Sci Bull MSTU CA Ser Comput Sci Appl Math (77):78–84
86. Pleshakova OA, Kozlov AI (2004) Making a management decision with a multi-alternative choice in conditions of different types of information parameters. Sci Bull MSTU CA (74):24–30
87. Eliseev BP (2015) Actual problems of training of specific specialists for the needs of civil aviation in the context of reforming of the educational system. In: Civil aviation: the past, present and future: materials of the international scientific and practical conference, pp 16–22
88. Eliseev BP, Koryagin ND (2016) A systematic approach to training managers of avia companies in the conditions of the information society. World Transp 14(3):244–253
89. Eliseev BP (2011) Methodology of system-technical management of professional education in civil aviation: abstract of the dissertation of the doctor of technical sciences. Ufa, p 32
90. Eliseev BP, Liu D, Zheleznaya IP (2015) The main indicators of the production activity of a typical air company. Sci Bull Moscow State Tech Univ Civ Aviat 221(11):153–161
91. Eliseev BP, Liu D, Zheleznaya IP, Wang B (2015) The methodology of decision-making, based on the results of the analysis of the air company's production activities. Sci Bull Moscow State Tech Univ Civ Aviat 222(12):170–174
92. Akchurin MR (2013) Improving the efficiency of management of aviation enterprise, based on improving methods, tools and models of aviation personnel management: abstract of the dissertation of the candidate of technical sciences. Moscow, p 16
93. Karelina IG, Sobolev AB, Sorokin SO (2015) Monitoring the activities of educational organizations—the initiative of systemic changes in higher education. Part 1. High Educ Today (6):37–46
94. Protasov TB (1989) Modular organization of the educational mathematical models. Development and application of expert systems: collection of scientific works. Science and Research Institute of High School, Moscow, pp 72–76
95. Beshelev SD, Gurvich FG (1973) Expert assessments. M.: Nauka, p 159
96. Box J, Jenkins G (1974) Analysis of time series. In: Pisarenko VF (ed) Forecast and management (1). M: Mir, p 406
97. Chinyuchin YuM, Vorobyov VV, Gerasimova ED (2019) Modeling of the operability of aviation equipment objects for a long period of maintenance. Aviat Ind 2:10–15
98. Vorobiev VV, Kharlamov AS, Ianchugin SJ, Ganiev ShF (2019) Influence of surface condition of taxiways and runways on time of aircraft departure. Sci Bull Moscow State Tech Univ Civ Aviat 22(3):8–15

Chapter 6
Designing of Innovative Components of Modernization of the Branch System of Education, Based on a Model-Oriented Approach and Evaluation of the Effectiveness of Their Implementation, When Training of the Personnel, in the Area of Maintenance of Air Transport

The innovative components, that implement the methodology of the model-oriented approach in the branch system of education, primarily relate to the implementation of modernization measures to restructure the educational cluster of civil aviation. In this case, there is a need to use an expert procedure for choosing a rational solution option set out in clause 2.2.

The problem orientation of this procedure consists in taking into account the options of modernization measures for the designing of the educational system for a given calendar period, synchronized with its current functioning. This requires, first of all, the systematization of many useful results and services, achieved during the introduction of innovative components of modernization, and the formation of a model of hierarchical management of the dual environment of functioning and development.

The final phase of structural transformations is the optimization of the elements of the content of training in the field of civil aviation. The model in this case is a tree of goals that combines the levels that characterize the content side of the implementation of educational standards of new generation. Optimization with using this model begins with the upper levels that determine the curriculum and the content of the disciplines. A distinctive feature of the application of the model-oriented approach is the description of the components of the lower levels, based on a reduced set of models, showing professional activity, which at the last level is transformed into components of the educational resources of the upper levels.

The analysis of the implementation of the designed innovative components allows us to evaluate, on the basis of monitoring information for 2014–2016 years, the effectiveness of using a model-oriented approach to improve the training of specialists in the field of civil aviation.

A. Borzova et al., *Modeling and Forecasting of Staffing in Civil Aviation*,
Springer Aerospace Technology, https://doi.org/10.1007/978-981-19-6238-7_6

6.1 Modeling and Decision-Making in Case of Modernization of the Branch Educational Cluster of Civil Aviation

The branch educational system in the area of air transport maintenance includes educational institutions of civil aviation of the Russian Federation, which include separate structural divisions (branches), that provide training for both secondary vocational and higher education programs.

The educational process is a continuous vertically integrated training in 8 enlarged groups of directions (specialties), which take into account the profile of training specialists to ensure the functioning of the air transport maintenance industry.

Modern requirements for training personnel in the field of civil aviation have set the task of implementing a set of modernization measures, aimed at ensuring of corresponding of the educational process to the needs of the field of air transport maintenance:

- structural reorganization of the network of branches of basic universities with the addition of secondary vocational education institutions;
- formation of an integrated system of higher and professional education;
- building an educational process, based on combining the requirements of educational and professional standards with the embedding of practice-oriented blocks (technological, operational and regulatory), which are agreed with employers and take into account the location and technical equipment that are necessary for the professional training of aviation specialists;
- implementation of a model-based approach to predict the release of personnel and their subsequent employment;
- creation of model-oriented educational resources that show the basic elements of theoretical and practical training of specialists in the places of their future employment.

For effective management of the modernization process, synchronized with the process of current functioning, we will highlight the basic elements of the structure of the educational cluster [1, 2], which change during modernization, taking into account the influence of external factors (Fig. 6.1): the educational process as an object of management and the organizational structure of the cluster as a system of management.

In the process of managing, the results of comparing tasks and areas of activity $x(t_0, t)$ with information about the level of achievement of goal $y(t_0, t)$ and the existing state of the managed object, determined by the modernization process, are solutions [3, 4]. The control center makes it possible to redistribute the resources $u(t_0, t)$ by its solutions, which synchronize the modernization and functioning process to the maximum extent, while compensating for certain inconsistencies between $y(t_0, t)$ and $\mathrm{x}(t_0, t)$, i.e., the influence of external factors, as well as internal influences, determined by the activity of elements in the system operation process, is determined.

Operational management has the following form:

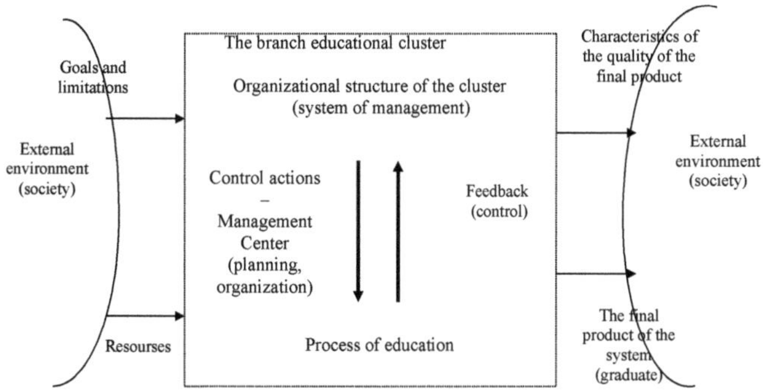

Fig. 6.1 Model of managing the functioning of an educational cluster

$$\Psi = \min[\Phi(x^e_{[t_0,t_1]}, y_{[t_0,t_1]}, u_{[t_0,t_1]}, \upsilon_{[t_0,t_1]})\big|y(t) = L(x_{[t_0,t]}.n^1_{[t_0,t]}); \quad (6.1)$$

$$\begin{aligned} &x(t) = F(x^0, t, u_{[t_0,t]}, \upsilon_{[t_0,t]}); u(t) = Q(x^e_{[t_0,t_1]}, y_{[t_0,t]}, \upsilon_{[t_0,t]}, t); \\ &\upsilon(t) = N(\upsilon_{[t_0,t]}, n^2_{[t_0,t]}); x^0, x^e \in R[x]; t \in [t_0, t_1], \end{aligned} \quad (6.2)$$

ψ is set of indicators of the quality of operational management, while the minimization operation is defined as reducing the costs of managing at certain points;

F, L and *N* are informal operators (algorithms) that determine the necessary and current information about the managed object, as well as about the external environment;

Q—an informal operator (algorithm) that allows you to make a choice or redistribute resources *u(t)*, in order to compensate for interference n^1 of perturbations n^2.

To describe the process of modernization in time of an educational cluster, the concept of development *{*R_t*}*, which is a purposeful temporary change of the structure *{S}* and function *{F}* of the cluster and the set $C = \{S;\ F;\ R_t\}$ for the purpose of modeling the structure, temporary functioning and development of an educational cluster are formed [5, 6].

The used components are considered from the point of view of the quality and efficiency of the system development.

The resulting set of optimal results should be divided into 3 different groups.

Each of the obtained groups of optimal results is characterized by various sets that have temporary changes. In the case, when the initial sets of results (*A*) will be ahead of the intermediate sets of results (*B*), then in the case ($A > B$) there is insufficient modernization activity of the managed party in the educational cluster.

Thus, the modernization regime, which is synchronized with the functioning and meets the most complete satisfaction of public needs, is expressed as: $C > B > A$, which determines the changes in the results of *C*, *B*, *A*. The systematization of the set of optimal results and services is presented in Table 6.1.

Table 6.1 Systematization of the set of optimal results and services

#	A set of optimal results and provided services	Description
1	Initial ones (*A*) that are not produced, but are consumed by the system	The system's perception of the conditions of the educational cluster's activity: material, energy and information
2	Intermediate ones (*B*), which are consumed and produced by the system	The product of the results of the activities of the internal elements of the educational cluster
3	Final (*C*), not consumed, but produced by the system	Elements of the results of the functioning of the educational cluster or its impact on the external environment, consisting in the release of specialists

The function that determines the planning of the process of functioning of the educational complex is to take into account the restrictions in the interaction of educational departments of an educational organization.

Each educational unit can influence the final result of the system by:

- finding information links, as well as hierarchical relationships between the structural units of an educational organization, that is, taking into account changes of the efficiency indicator of the system for management them;
- finding an approach to overcoming the limitations that are imposed on these connections.

The management system for the functioning of an educational cluster includes:

- the investment sphere, which determines the methodology and organization of processes—study ones, pedagogical ones and scientific ones, as well as determining the manifestation of independence, creative activity, associated with the development of the cluster—the formation and decision-making in conditions of complete uncertainty (scientific and research sector, scientific and research laboratories, etc.);
- infrastructure sphere—processes that provide management of information, material and energy flows in an educational cluster. The activity of study units is determined by the development of restrictions on the relations and connections between the components under consideration as a management system, in particular—personnel management, planning and financial management, mechanical and energy services, logistics;
- the administrative sphere, which includes elements that do not affect the composition and volume of the final results. The functioning of these links is determined by the conservation of the specified working conditions of the educational complex (office, study management, library, etc.);
- the sphere of production, which includes elements that implement, what is defined in the investment sphere, allocated in the infrastructure sphere and supported in the administrative sphere, i.e., is the educational process.

Table 6.2 Modes of development of the educational system

Dynamic relations by sets	Description of the work of the educational cluster
$C > B > A$	The educational cluster operates in a basic or similar mode: The pace of creative search for the creation of the final product is ahead of the growth rate of educational technologies and career guidance work
$C > A > B$	The need to review the activities of the infrastructure and administrative areas of the educational units of the educational cluster (deviation from the basic version)
$B > C > A$	The infrastructure sector is ahead of the investment one. The need to intensify scientific developments and research on modern and important problems of higher education and changes in the intra-university work schedule
$B > A > C$	The investment sphere does not reach the infrastructure and administrative ones. Scientific developments and research on the issues of the educational cluster are not carried out
$A > C > B$	The infrastructure sphere of the educational complex does not reach the level of the teaching sphere
$A > B > C$	There is a need to pay the main attention to scientific research and development and to improve the interaction between all departments of the educational complex that carry out the study process

The presented types of activities are interconnected with each other as stages of the life cycle of the educational system: from scientific research, further, making a choice of restrictions on relationships and connections and maintaining certain conditions, ending with the process of personnel graduation.

The educational cluster functions in 6 possible modes, as a dynamic relationship between the sets C, B and A (Table 6.2).

When managing the modernization process, which is synchronized with the activities and development of the educational system, the classification of certain goals is used:

- the process of forecasting the results of activities, taking into account the interdependence of decision-making conditions and information support, perceived as a chain: from the goal → the problems → tasks → before the task;
- a certain goal consists in the following sets: from purposeful problems → a set of problem-oriented tasks → to a certain number of oriented task.

In this case, the constructed tree of goals provides the ability to group the main activities of the system with different degrees of detail. 4 basic functions are defined in the functioning of the educational system: from functioning, development, marketing, to management (Fig. 6.2) [7].

Using the details of certain goals and the following regulation of the work of the elements of the educational cluster, a model of step-by-step management of the process of modernization of the educational system is formed (Fig. 6.3) [8].

This system has a control center and n active blocks. So the characteristics of the ith element (objective function, possible states, etc.) are determined by the parameters

operation

– evaluation of study work (questionnaire «teacher through the eyes of students», test control;
– support of study work (study and methodological activities);

development

– introduction of a modern study environment (study and methodological support, equipment);
- new pedagogical technologies (training technologies, based on new equipment, diagnostics of the study process);
– relevant areas of training (specialties, profiles, disciplines);
– pedagogical researches

marketing

– formation of the need for educational services;
– «sale» of specialists;
– commercial activity (technologies, software);
– image and achievements of the university (Public Relations)

management

– organization of new workplaces;
– organizational and methodological activities

Fig. 6.2 Basic functions of the university's educational system

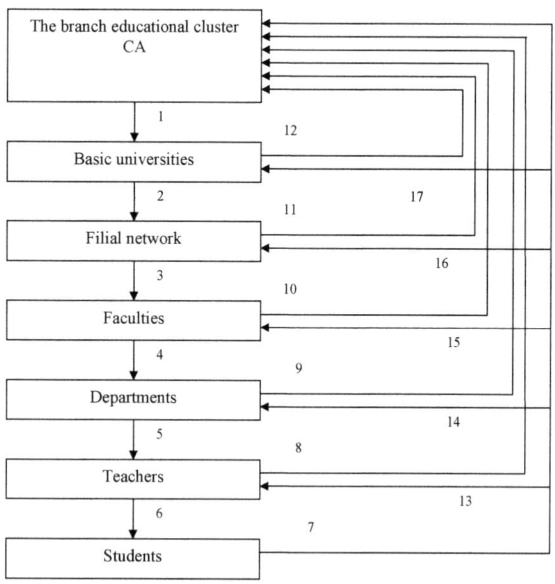

Fig. 6.3 Structure of step-by-step management of the process of modernization of the educational cluster

(staffing of the teaching staff, the formation of the contingent, the content of education, learning equipment, etc.) $r_i \in \Omega_i$, $i = \overline{1, n}$. For example, for a basic university, the center is the rector's office, for a direction/specialty the center is a department that graduates of specialists. Each considered parameter r_i is characterized by an estimate of s_i. The activity of the desired system occurs in stages [9]:

- when collecting information, the elements transfer to the center of evaluations $\{s_i\}$ of parameters $\{r_i\}$;
- the center assigns plans $x_i = \pi_i(s)$ to the elements $i = \overline{1, n}$ in the planning process, and using the formed estimates and taking into account the planning procedure: $\pi\colon \Omega \to X$, where $\Omega = \prod_{i=1}^{n} \Omega_i$, $X = \prod_{i=1}^{n} X_i$ is a set of possible plans that include the areas of activities of the educational cluster in order to improve the efficiency of functioning, as well as the quality of the learning process;
- when choosing a strategy, taking into account planned tasks, the elements determine for themselves the states $y_i \in A_i(r_i)$.

The interests for the elements and the center are determined by the functions of goal $\Phi(x,y)$ and $f_i(x_i,y_i)$, $i = \overline{1, n}$, where $x = (x_1,\ldots,x_n)$ and $y = (y_1,\ldots,y_n)$. Thus, for each element is defined the choice of the state and the message of estimations. Taking into account the fixed nature of the plans, as well as the rational behavior of the elements, the actions that are selected will allow to maximize the functions of goal, i.e.,

$$y_i^* \in \arg\max f_i(x_i,\ y_i) \tag{6.3}$$

$$y_i \in A_i(r_i)$$

Let $\psi\ (x,r) = \Phi(x,y^*)$. Then, the value of the function of goal of the element under consideration will depend on the plans that are assigned and depend, in own turn, from the estimates, reported by the elements. So, the element under consideration itself will try to present an estimate that will allow to maximize the function of goal, taking into account, that this estimate may not coincide with the correct one. In this case, the communication of reliable information in the problem under consideration is an equilibrium strategy for the components under consideration.

Then, a task is defined for the center—the definition of a planning procedure that allows to ensure in the solution (at the equilibrium point) that the value of its objective function is as much as possible.

An important characteristic of the mechanism is its effectiveness:

$$\mathrm{K}(\Sigma) = \min_{r\in\Omega} \frac{\min_{s\in P_s} \Psi(x, r)}{\Psi_B(r)} \tag{6.4}$$

where $\psi_B\ (r)$—"weight" function and P_s —a set of equilibrium strategies of $(s_* \in \Omega)$ elements.

$X_i\,(s_{\text{-}i})$ is given for each set. The procedure of planning of modernization measures is considered:

$$(x,\ s) \to \max \tag{6.5}$$

$$x \in X$$

$$j_i(x_i,\ s_i) = \max\ j_i(z,\ s_i) \tag{6.6}$$

$$z \in X_i(s_{-i})$$

where $\varphi_\text{i}\ (\cdot)$—the function of preference for the i-th of the considered active element.

Then, the procedures (6.5) and (6.6) are an open control mechanism. In this case, the condition (6.6) is a condition of perfect agreement.

The management of the modernization process along the vertical 1-2-3-4-5-6 is carried out taking into account the prognostic assessments, discussed in chapter 3 and feedback, which is based on a comparison of indicators for monitoring the effectiveness of the functioning of universities in the standard operation mode and after the inclusion of modernization blocks [10, 11].

The block diagram of the modernization process and decision-making in the case of the implementation of hierarchical management is shown in Fig. 6.4.

when choosing modernization measures, based on the conducted expert-monitoring analysis

Module 1 shows the information support that characterizes the elements of hierarchical management, as well as the relationships between them, shown in Fig. 6.3.

Module 2 allows the control center to implement options $W_1,\ldots,W_1,\ldots,W_L$ of the composition of modernization measures for a certain calendar period [12, 13].

Modules 3, 4 and 5 support the classification procedure for the properties of the educational cluster for the current activity mode and also when forming modernization measures for the next calendar period.

The selection of the more important ones, which represent universal indicators of quality $y_g,\ g = \overline{1, G}$, the values of which are determined by the dependence from the initial parameters of functioning and from the variant of modernization measures is carried out.

Modules 6 and 7 affect the bringing of all indicators of quality $y_g,\ g = \overline{1, G}$ to scale [0,*A*] (mostly—[0,1]) taking into account a certain method of rationing. It is possible to apply rationing $\hat{y}_g,\ g = \overline{1, G}$ by minimum and maximum values $y_g(w_l)$, $l = \overline{1, L}$, considered in [14].

Module 8 shows the definition of the structure of the model for the formation of an integral assessment of quality in the form of various multiplicative and additive convolutions. In the book, the definition of integral estimation F_l for each considered

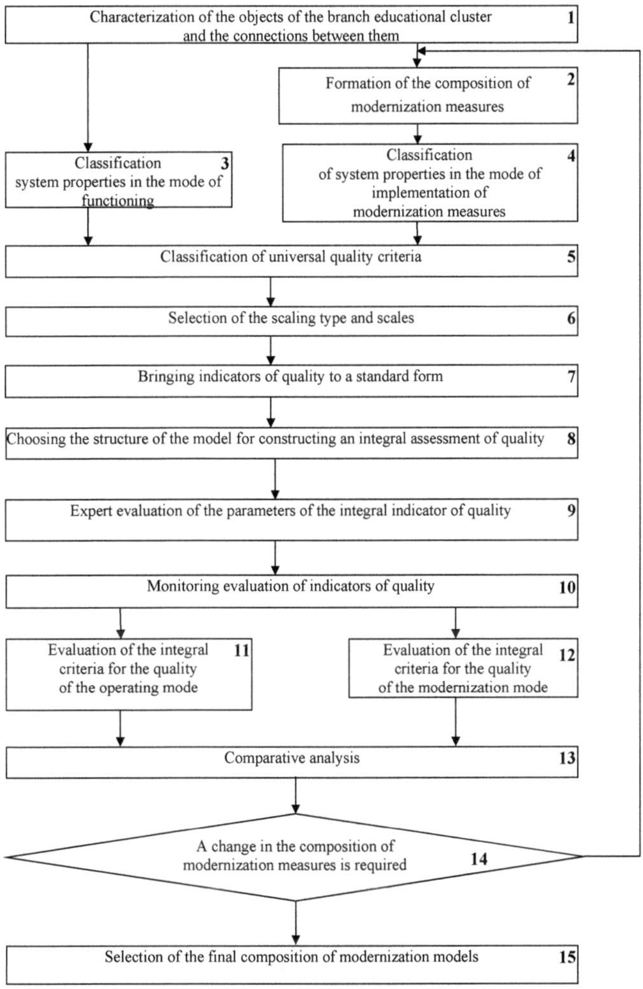

Fig. 6.4 Structural diagram of the modeling and decision-making processes

option $w_l, l = \overline{1, L}$ by the expression of the average weighted convolution, which is given in [14], with the determination of the module parameters is considered.

Modules 10–15 are combined into one expert-monitoring procedure, taking into account the adaptive approach, which is presented in detail in [14, 15]. An important requirement is to show the expert during the survey not only the value of the indicators, that are monitored $y_g(w_l^k)$ [16] for the variant of modernization measures, but also the values of the indicators y_g^φ, $g = \overline{1, G}$, used for the mode of operation being carried out. Expert usually takes into account the values y_g^φ, $g = \overline{1, G}$ in the following cases:

- when determining the most non-acceptable indicator from the available set $y_g(w_l^k)$, $g = \overline{1, G}$, which includes a worse value than in the mode of ongoing operation;
- when selecting the gradation of a linguistic variable;
- in the case of determining the final solution in the process of forming a set l_2, based on the conducted comparison of the values of integral estimates F_l for the considered variant of modernization measures and F^{φ} for the mode of current functioning.

Thus, the model-oriented approach makes it possible to choose step by step a variant of modernization measures for each new period (calendar), on the one hand synchronized with the current functioning and on the other hand to allow realize the innovative development of the CA educational system.

6.2 Optimization of Components of the Training Content in the Field of Air Transport Maintenance, Based on Expert Analysis With A Focus on A Model-Oriented Approach

The basic principle of the organization of the personnel training system in the area of air transport maintenance is the educational process, formed taking into account the gradual transition from educational and cognitive to professional activity. A systematic consideration of the approach to training specialists is necessary in the conditions of increasing modern requirements for the quality of training—training of specialists, correction of the goals as well as the content of training [17–21].

The problems of forming the optimal structure of the educational process lead to the requirement to develop a model of the content of training specialists, which is a complex structure that includes elements of the system, as well as bilateral and multi-lateral relations among them.

The content of training is formed in the form of a hierarchical multi-level system, and the main task of which is to assess the overlying levels, taking into account the interaction of the hierarchy levels, and not the interdependence of the elements, located at these levels [22, 23]. The objects, that are evaluated, are considered as blocks of the same level of the hierarchy, which is the tree of goals (Fig. 6.5).

The curriculum for areas or specialties in the field of civil aviation is considered as the zero level, and the basic study disciplines are considered as the first level. A characteristic feature of using the model-oriented approach is served the description of the elements of the 3rd level, based on the reduced set of models of professional activity μ_n, $n = \overline{1, N}$ [24], and the fourth models that are transformed into components of educational programs O_v, $v = \overline{1, V}$, forming the funds of available educational resources.

An element of the tree of goals has a certain coefficient of relative importance, which determines the indicator of the contribution of this element to the achievement

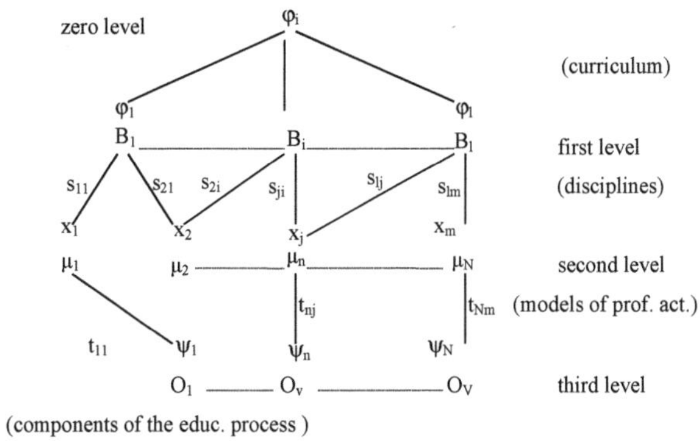

Fig. 6.5 Tree of goals for developing the content of training in areas and specialties in the field of civil aviation

of the main goal for the system. Thus, the coefficients of the relative importance of the considered goals belonging to one level of the considered tree of goals are the relative weights of the elements of this level, according to the qualitative sign of importance (significance). The coefficient of relative importance of the main goal is the training of specialists, $A = 1$. Since the achievement of the main goal is uniquely determined by the achievement of the first level goals by the main subsystems of this system, the values of the coefficients of relative importance of the elements, located at the first level (discipline) $\varphi_1,\ldots,\varphi_1$, which are determined by experts, are determined first, and the condition $\sum_{i=1}^{l} \varphi_i = 1$ must be met.

At the next stage, the values of the relative weights of the connections of those elements, that are located on the second level with the elements of the first level of S_{ji}, are calculated, while $\sum_{i=1}^{m} s_{ij} = 1$. So, requirements are grouped into groups (clusters) in order to determine the priorities of all elements, when grouping—this is more efficient than the process of common processing of all elements. First, the requirements are integrated, taking into account the compliance with various types of professional activities of graduates, and further, each cluster is ranked by importance, taking into account the elements that are at the previous level.

At the next stage, the adjacency matrix $A = \|a_{ij}\|,\ i = \overline{1,m},\ \ j = \overline{1,m}$, is constructed, which establishes all logical connections between the requirements, taking into account the type of activity of specialists

$$a_{ij} = \begin{cases} 1, & \text{if } a_i \text{ more preferable than } a_j \\ 0, & \text{in otherwise} \end{cases} \tag{6.7}$$

At the next step, experts fill in the matrix, using the method of paired comparisons. Taking into account its large dimension, initially only those requirements are highlighted, the importance of which can be clearly defined, which allows reducing the number of performed operations.

For each pair of objects *i*, *j* under consideration, an evaluation is conducted.

$$a_{ij} = \frac{s_i}{s_j} \tag{6.8}$$

where s_i—weight of object *i* and by matrix *A*. Considering the normalizations $s_1 + \ldots + s_m = 1$, further the eigenvector of the desired matrix *A* is determined, which, after normalization, is a vector of priorities $S = (s_1,\ldots,s_m)$.

The eigenvector is determined, using the statement—the eigenvector, which corresponds to $\lambda_{\max}$, is the normalized sums of the row elements of the considered limit matrix in accuracy the *k*th degree A^k of the matrix *A*.

$$\lim A^k \times e = c \times s, \quad ||A^k|| = e^T \times A \times e \tag{6.9}$$

$$k \to \infty ||A^k||$$

where *c* is constant, $e = (1,\ldots,1)^T$, *s* is the eigenvector that corresponds to $\lambda_{\max}$ and $\lambda_{\max}$ is a maximum eigenvalue that shows the assessment of consistency, showing the proportionality of preferences.

The calculation procedure is completed, when the difference between these sums, obtained in two consecutive calculations, is less than the a priori given value.

Next, group estimates of the necessary coefficients of the relative importance of the requirements are calculated, taking into account that the assessment is carried out by a team of experts.

$$s_{ij} = \left(\frac{1}{h}\right) s_{ij} \tag{6.10}$$

where *h* is the number of experts.

Then, the coefficients of relative importance for the elements of the 2nd level are determined.

$$xj = \sum_{i=1}^{l} \varphi_i * s_{ij} \tag{6.11}$$

on condition $\sum_{i=1}^{m} x_j = 1$.

At the next stage, the relative weights of the links for the elements of the 3rd level with the elements of the 2nd level t_{kj} in case of $\sum_{k=1}^{n} t_{kj} = 1$ are determined.

Then, the next level blocks are grouped into clusters, using the method of paired comparisons, which allows you to determine the more preferred element in each pair under consideration, but also to show the degrees of preference of objects.

The result of performing paired comparisons of n-objects is represented as an oriented graph, containing n vertices, in this case, each of them corresponds to a specific object. So, if the object a_k is more preferable than the object a_r, according to the specific feature under consideration, then there is an arc in the existing graph that is directed from the vertex a_k to the vertex a_j. The formed graph does not contain loops, and then, each pair of vertices will be connected by one and only one arc.

The choice, based on a generalized qualitative feature, is carried out as a choice, based on a combination of several available simpler features. In the process of comparing one pair of objects, the expert determines one specific feature as decisive, for another pair of objects—another specific feature is more important, which leads to contradictory judgments.

The results of comparing all available pairs of objects are further shown in the form of a relationship matrix $A = \|a_{\text{ru}}\|$.

If the expert's preferences are step by step and the graph does not contain contours, then the sums of elements, located on the corresponding rows of the matrix A, are used as weights of objects.

Relative weights of objects.

$$t_u|_{u=kj} = \frac{\sum_{r=1}^{P} a_{\text{ru}}}{\sum_{r=1}^{P}\sum_{k=1}^{P} a_{ru}} \tag{6.12}$$

Taking into account that the assessment is made by a certain group of experts, the joint assessment of the obtained relative weights of objects can be formed as the arithmetic mean of the estimates of experts:

$$t_u = \frac{1}{H}\sum_{h=1}^{H} t_{\text{uh}} \tag{6.13}$$

Then, taking into account the coefficients of relative importance for the elements of the second level, i.e., models of professional activity, the coefficients of relative importance for the elements of the 3rd level, i.e., the components of the educational process, are calculated.

$$\Psi_k = \sum_{j=1}^{m} x_j * t_{kj} \tag{6.14}$$

in the case of $\sum_{k=1}^{n} \Psi_k = 1$.

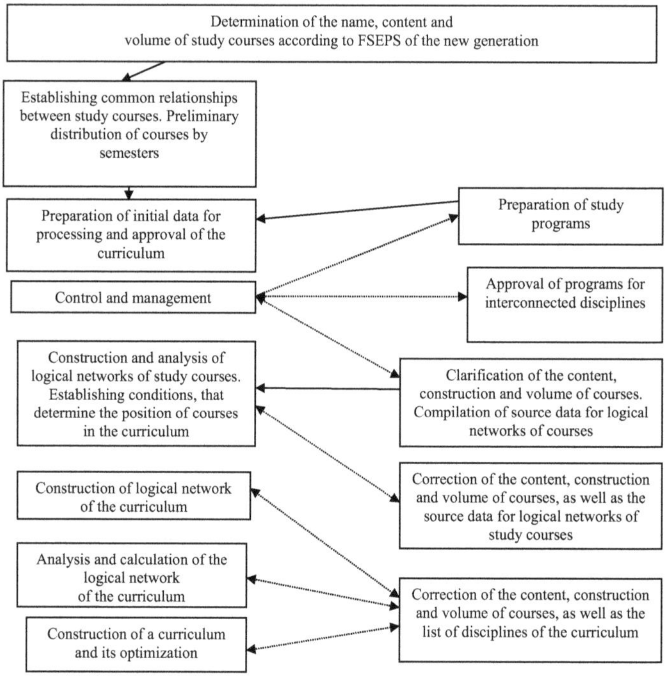

Fig. 6.6 Block diagram of the formation of the curriculum

So, on the basis of the formed tree of goals that builds the content of training, the basic components of the educational process are developed.

The process of developing the upper levels of the tree of goals of content of study (Fig. 6.6) includes three stages:

1. Taking into account the educational standards of the new generation (FSEPS 3 ++) [25], the content and volume of study disciplines are determined.

   So, the requirements for general engineering and general scientific disciplines are sequentially formed, which represent groups of main issues for study, and the disciplines themselves. The resulting sequence determines the fulfillment of all the most important requirements for the process of training a specialist as a whole, as well as the preparation of the required base for studying each discipline.

   At the next step, the general relationships between all disciplines of the curriculum are established, i.e., which courses need to be agreed on the content, methodological content and time of study.
2. Development of the content and educational and methodological support for each course and determination of its place in the curriculum.

   Next, a graph of the training course is formed, which shows its relationship with other disciplines on topics and allows you to assess objectively the roles and places of this course in the structure of training specialists.
3. Development and optimization of the curriculum structure.

At this step, an important one for solving the problem of developing an optimal curriculum is an oriented graph of the educational process, for which the vertices correspond to the elements of the curriculum, and the arcs correspond to the connections between them. At the same time, the intensities of these connections q_{st} are calculated by experts in the process of filling in the corresponding relationship matrix. The problem to be solved is to divide the graph of the educational process into layers, taking into account the conditions, when the vertices of the same level should not be connected to each other ($q_{\text{st}} = 0$). At the same time, the number of layers in this case should be accordingly.

Taking into account that the relationship matrix has a sufficient size, it should be divided into submatrices in accordance with the blocks of the curriculum. Then, you should choose a rational ordering of the elements of the curriculum under consideration, when forming the relationship matrix, then at the first all the elements of one block are listed, and then the elements of the next block are listed, etc. It is advisable to arrange the elements of each block in the sequence in which they should be studied as far as possible, when compiling the relationship matrix.

The general interrelations are also determined, which consist in determining which courses need to be further coordinated with each other in terms of content and sequence of study (a matrix is also being constructed). Each course in the matrix corresponds to a selected row and column, and the number of which corresponds to the numbering of the course according to the set list. The specified unit at the intersection of the row and column shows that the discipline from the nth column should be based on the course from the nth row. In this case, each column shows all the courses, the information generated in which should be used, when studying the discipline corresponding to the column.

The presentation in the matrix in some generalized form of the need for the formation of information by other courses in the interests of a certain discipline, as well as the intended consumers of information that should be formed in this course under consideration, makes it possible to initially plan the division of all study courses into groups that correspond to a given sequence of mastering them.

Combining disciplines into some groups is carried out by the method of matrix reduction, which consists in identifying zero columns that correspond to study courses that do not require any additional information from other disciplines to start studying, the study of which can be carried out from the beginning of the study process. Then, by excluding the rows, corresponding to the courses of the first group, the original matrix is reduced. So, in the truncated matrix, the zero columns are defined further, as well as those disciplines that belong to the second group, etc.

The submatrices, that correspond to the blocks of the curriculum, are located on the main diagonal of the matrix (Fig. 6.7).

After the development stage of the relationship matrix of the elements of the educational process, the constructed graph is divided into levels. So, the set of vertices then is divided into a disjoint subset, ordered so, that, if the vertex in question belongs to a layer with the number n and is the beginning of some considered arc of the graph, then the final vertex of this considered arc will belong to a layer with the number greater than n. Then, the number of layers corresponds to the number of semesters.

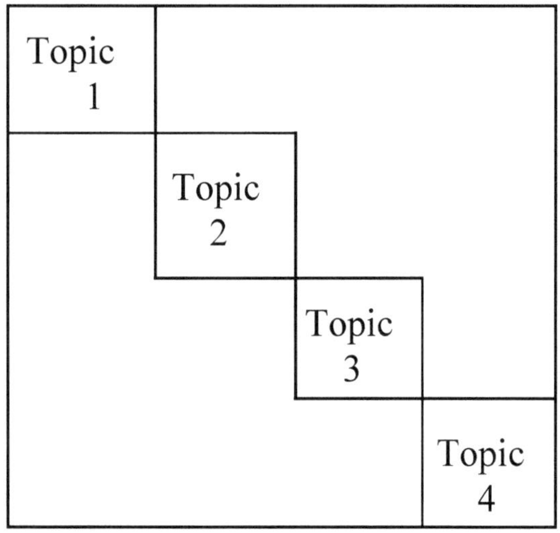

Fig. 6.7 Location of the submatrices of the relationship matrix

Thus, a set of vertices, corresponding to the number of semesters, is formed. At the same time, the number of elements in each of these considered sets is not equal, in general, to the specified volume of the semester load. Such a task is to redistribute elements between these sets, which will ensure the required loading of semesters.

The determining requirements are the logic and consistency of the learning process, which show that the study of this material is based on what was previously studied. The lack of a clear sequence in the learning process determines a decrease in its quality and irrational spending of educational time. At the same time, almost identical questions are often found in different disciplines; each of them should be thoroughly considered once in the discipline that was studied earlier and then only mentioned in other disciplines.

The elements of educational programs O_n, $n = \overline{1, N}$, which are formed on the basis of professional activity models, are considered further as educational units that need to be placed in an optimal sequence, taking into account interactions with the zero and the first layer of the tree of goals of content of the learning (by areas, specialties) in the field of air transport maintenance.

The existing study elements, as well as their logical relationships, should be presented in the form of a graph of logical relationships, as well as the corresponding matrices of study elements [26].

The matrix of formed logical relationships of n study components $B = (b_{ij})$ shows the number of logical relationships b_{ij}, established between i and j study elements $\left(i, j = \overline{1, n}\right)$. The sequence of learning of study components during a certain period of study should be represented as a sequence $\pi = (i_1, i_2, \ldots, i_k, \ldots, i_n)$, where i_k is the number of the available study unit that occupies the kth place on the left $\left(k = \overline{1, n}\right)$ in the sequence π (Fig. 6.8).

Then, the algorithm for determining the optimal sequence π includes:

1. Definition of the matrix of logical relationships of study components

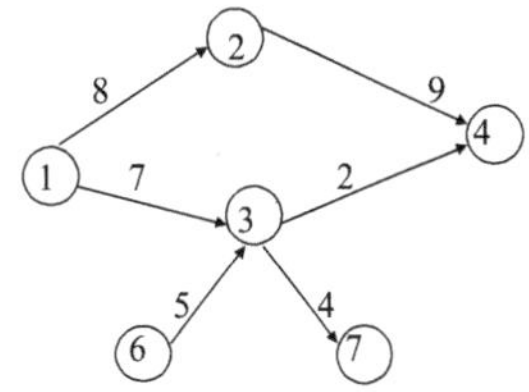

Fig. 6.8 Graph structure of logical relationships of study elements

$$B = (b_{ij}) \tag{6.15}$$

where b_{ij}—the number of specific logical relationships that are directed from the basic study—*i*, to the some learning study element—*j*.

2. Allocation of the amount of time, allocated for the learning of study components, which are set by the vector of volume—T.

$$T(t_1,\ t_2,\ \ldots,\ t_k, \ldots,\ t_n) \tag{6.16}$$

where t_k is volume in hours of the study element—*k*.

3. Determining the predicted numbers of study elements K_i, (vertices of the graph)

$$k_i = \sum_{l=1}^{n} b_{il} - \sum_{p=1}^{n} b_{pi} \tag{6.17}$$

where. $i = \overline{1, n}$

4. The function $F(\pi)$ is represented as

$$F(\pi) = \sum_{i=1}^{n} (k_1 + k_2 + \ldots + K_i)t_i \tag{6.18}$$

Function $F(\pi)$ is the total number of breaks of the formed logical relationships between all the study components of the specialty/direction of training, which are determined for the entire study cycle, taking into account the period of learning of each study component, as well as taking into account the relations of precedence, according to logical relationships.

So, the sequence π, at which the lowest value of $F(\pi)$ is reached, is determined at the following stages.

5. Formation of matrix $A = (a_{ij})$, where each element is defined by the following rule:

$$a_{ij} = k_i^* t_j - k_j^* t_i \tag{6.19}$$

where K_i is taken from (6.10) *and* t_i is taken from (6.16),.$i = \overline{1, n}$

Then, the minimum value of the function $F(\pi)$ under consideration and also the minimum value of the function $S(\pi)$ must be achieved on the same sequence π under consideration.

$$S(\pi) = \sum_{i=1}^{n-1} \sum_{j=i+1}^{n} a_{ij} \tag{6.20}$$

So, the expression (6.18) defines a certain sum of the above diagonal elements of the matrix A under consideration.

Then, the number of the vertex of the graph under consideration should be identified with the number of the place in the sequence π under consideration or with the ordinal number of the column and row in the desired matrix A.

6. Taking into account the logical relationships that have arisen between all pairs of vertices of the graph of logical relationships of study components, the relations of precedence that are formed, when using the concept of direct and inverse transitive closure, are determined. In this case, the direct transitive closure A_j^- of vertex j is the set of vertices that can be reached by some path from vertex j. The inverse transitive closure A_j^- of vertex j is the set of vertices of the graph from which there are paths to vertex j.

The expression C_j.

$$\hat{\Gamma}_j \cup \hat{\Gamma}_j^- \tag{6.21}$$

is the set of vertices of graph, with which the considered vertex j is related by the relations of the order of precedence along any path. Then, the set of vertices of the graph is E.

Then, the expression.

$$E \backslash \hat{\Gamma}_j \cup j \tag{6.22}$$

is the set of study components, with which the considered study unit j is not connected by the existing outgoing logical relationships, both through other study elements and directly.

This expression.

$$E \backslash \hat{\Gamma}_j^- \cup j \tag{6.23}$$

is the set of study components, with which the considered study element j is not related by incoming logical relationships, both through other study elements and directly.

Then, the expression.

$$D_j = \left[E\backslash\hat{\Gamma}_j \cup j\right] \cap \left\lfloor E\backslash\hat{\Gamma}_j^- \cup j\right\rfloor = E\backslash\left(\hat{\Gamma}_j \cup \hat{\Gamma}_j^- \cup j\right) \tag{6.24}$$

is the set of study elements, with which the study element j is not related by logical relationships for the entire study cycle.

The set D_j allows us to compare with which study components in the educational process under consideration, and the study component j can be rearranged in time, without breaking logical relationships. This property is taken into account, when searching for the optimal sequence π for the function $F(\pi)$.

Then, the expression (6.24) determines the revealed disunion of the relationships of the considered study component with other components. In this case, when summing the value of the number of D_j elements for all n study components, an indicator is formed that characterizes the disunion of all the considered study components, according to the available logical relationships for the entire period of study in the specialty/direction.

Thus, the complete disunion of study elements will be characterized by the number$n(n-1)$.

Then, the following coefficient K_0 of filling of the matrix of logical relationships of the study components of the specialty/direction is introduced.

$$K_0 = 1 - \frac{\sum_{j=1}^{n} |D_j|}{n(n-1)} \tag{6.25}$$

where $|D_j|$ is number of elements of the set D_j.

The presented coefficient K_0 will determine the structural disunion of the study components, according to the main logical relationships. In the case, when $K_0 = 0$, the learning of the study components is carried out independently of each other, i.e., they are completely unrelated. Then, if $K_0 = 1$, then no study component is subject to rearrangement relative to other study components without breaking logical relationships that determine the uniquely determined ordering of study components that determines the optimal and only sequence of their learning.

7. Taking into account Eq. (6.22), pairs of vertices of the graph of logical relationships of all study components are determined, which are related by the relation of precedence " $\rightarrow$ ", both through other vertices of the graph, and directly.

The elements a_{ij} of the matrix A under consideration, corresponding to all the available pairs of vertices of the graph, in this case are formed over the main diagonal of this matrix A, and then, the total contribution from these components to the total (6.21) will be stable for all permissible permutations of columns and rows of the matrix A (6.21) or may be different for possible permissible sequences. In this case, the optimization of expression (6.21) is achieved due to the elements of the matrix A (6.20) under consideration, which will correspond to pairs of vertices of the considered graph of logical relationships of study components that are not related by relations of precedence, according to the obtained logical relationships both through other study components and directly.

Thus, the process of modernization to improve the system of training personnel in the field of civil aviation will be supplemented by an optimization choice of elements of the study content with subsequent orientation to a model-oriented approach, as well as educational standards of a new generation.

6.3 Analysis of the Effectiveness of Application of the Model-Oriented Approach in the Modernization of the Civil Aviation Branch Education Cluster

The developed model-oriented approach in the modernization of the branch educational cluster of civil aviation is aimed at implementing one of the tasks of the strategy for the development of the transport system of Russia until 2030 year—providing the industry with a sufficient number of highly professional specialists in the following areas:

- in the area of development and implementation of projects for the functioning and development of transport systems;
- in the area of maintenance of transport infrastructure, as well as vehicles, that are created in the process of implementing the strategy;
- in the area of providing transport and logistics services to the consumer, as well as other transport services;
- in the area of management of transport complex.

The modernization of educational institutions of civil aviation provided an opportunity to form an educational cluster, which allowed the concentration of intellectual and material resources into a single one as part of enlarged university complexes of both federal and regional significance, which have a wide network of territorial divisions (branches), that allow providing a full formed educational cycle, from the training of qualified workers and workers with secondary vocational education, as well as all types of continuous studying.

Three large centers have been formed in the cities of Moscow, St. Petersburg and Ulyanovsk with a developed branch network, including educational organizations, located in all federal districts and making it possible to provide highly qualified personnel to all Russian aviation enterprises that implement air transportation of civil assignment.

The presented model of hierarchical management of the modernization process of the educational cluster of civil aviation made it possible to consider the process of functioning and development of educational organizations, included in the cluster, through the use of hierarchical relations and information links between the elements of the educational process, to consider the functioning of the educational cluster in real time and, if necessary, make adjustments in case of deviation from its planned indicators.

The development of the educational cluster through research (fundamental, search and applied), primarily on the basis of large university complexes, strengthening their base (social, material and technical, scientific and laboratory), the formation of centers (research and production, innovation and implementation), technoparks, providing them with the latest samples of software, equipment, technologies, allowed during the annual monitoring of the activities of universities, conducted by the Ministry of education and science of the Russian Federation, increase the main indicators for the last 3 years by an average of 22.34%.

The developed model for the formation of educational content in the areas and specialties of training in the area of air transport maintenance made it possible to rationally define the goals of education and provide targeted management of the content of problem-oriented educational programs, based on combining the requirements of Federal state educational standards of higher and secondary vocational education, professional standards, as well as the requirements of employers, when training specialists.

Improving the quality of training of specialists in educational institutions of civil aviation has allowed to increase the number of aviation personnel, employed at civil aviation enterprises by 11.2% over the past 3 years.

Thus, the application of a model-oriented approach in the modernization of the branch educational cluster of civil aviation and the improvement of the personnel training program in accordance with the ongoing changes in market requirements has made it possible to increase the effectiveness of the formation of personnel policy in the area of air transport maintenance.

6.4 Ways to Improve the Quality of Pre-professional Education on the Example of the Moscow State Technical University of Civil Aviation

The effective system of training personnel for civil aviation at the university is largely influenced by the competent organization of pre-professional training.

Pre-professional training of schoolchildren is one of the main conditions for the correct choice of a university and, as a result, their future profession. The project "University Saturdays", organized by the Department of education of the City of Moscow has been one of the main links in the implementation of this task for many years. During this time, hundreds of thousands of schoolchildren, their teachers and parents, as well as thousands of the most competent, dynamic and creative teachers of Moscow universities have already become participants in this project. In this section of the book, using the example of one university—the Moscow state technical university of civil aviation (MSTU CA), the structure of the project organization will be shown, as well as an assessment of its sustainability and quality.

The main components of the “University Saturdays” project are:

(1) The Department of education of c. Moscow. The functions of the Department include providing a grant to the university for organizing and conducting events within the framework of the project, receiving reporting materials, organizing electronic registration on the website of schoolchildren, students of colleges and their parents, as well as monitoring the quality of the project within the university.
(2) University. Its functions include organizing an application for a grant for carrying out activities within the framework of the project to the Department of education. In case of receiving a grant, the University provides several tens events in the period from March to December. The events consist of various popular science lectures, workshops, excursions, master classes and trainings, which are organized mainly in the main building of the University, as well as in the Educational aviation and technical center of the university in Sheremetyevo. The university allocates classrooms and specialized centers for organizing events, organizes the passage and registration of participants, who have arrived at the event, and also actively participates in the agitation of schoolchildren from various schools and other secondary educational institutions for participation in the project events.
(3) Responsible for the organization of the entire project at the University. This person is appointed from among the members of the teaching staff. Its functions include the organization of lecturers, as well as those responsible for conducting workshops, master classes, trainings and excursions to participate in the project. After that, it is necessary to collect the topics of the events in a short time and select the most relevant and interesting among them. Further, the person responsible for organizing the “University Saturdays” project should regularly contact the Department of education of c. Moscow on the following issues:
(a) revision of the application topics, as well as their descriptions for the best adaptation to the perception of schoolchildren;
(b) notification about the date of the events, as well as the date of the beginning of their registration on the website;
(c) timely notification of changes in the schedule (replacement of the lecturer, events, postponement of the event to another time, etc.);
(d) meeting schoolchildren at each event, providing background information before the event.
(4) Schoolchildren, students of colleges, their teachers and parents. In addition to participating in the events of the University Saturdays project, their functions include informing the person, responsible for organizing the project about the evaluation of the event attended, as well as the Moscow Department of education. On the basis of this information, corrective information is developed, and on the basis of which, the effectiveness of the entire cycle of measures can be increased.

In Fig. 6.9, the structural scheme of the project “University Saturdays” is given.

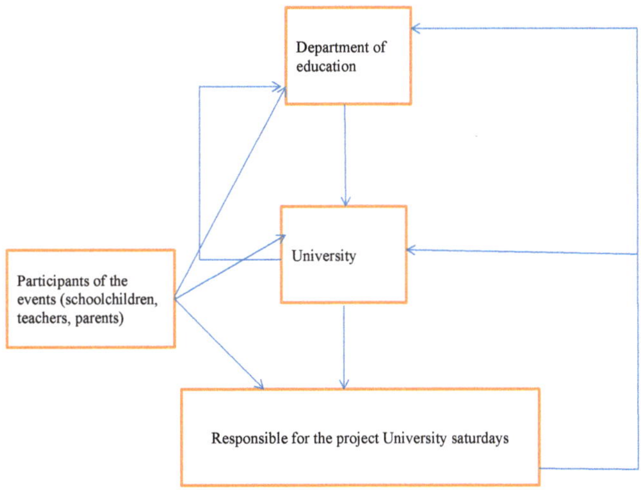

Fig. 6.9 Block diagram of the project "University Saturdays"

Here are some comments on this scheme. The stability of the entire project organization system is ensured by the presence of direct and feedback links both within the university among all persons, participating in the project, and on the part of the Moscow Department of education, as well as schoolchildren, students of colleges, their teachers and parents [27, 28].

For example, a few years ago there was a case of the need to transfer one of the events to another day. The person, responsible for the project, informed all the prospective project participants, who registered on the site in time, but did not inform the Department about it, deciding, that the problem had already been solved. As a result, no changes were made to the website and some of the participants, who did not register electronically and, not knowing that the event was transferred, wasted their time on the weekend.

Or another example. To conduct events within the framework of the project at the University, the person, responsible for organizing the project, must constantly coordinate efforts with various structures and services: the management of university to solve many urgent organizational issues; the department of pedagogical work to interact with schools in order to agitate schoolchildren to register on the website and participate in events; heads of departments to approve the participation of their teachers in holding events; the study department for allocating classrooms and providing them with the necessary modern technical learning tools; administrative and household management for solving a significant number of issues, related to ensuring the passage of participants to the university building, as well as their registration.

Here are some indicators that characterize the effectiveness of the "University Saturdays" project at the University [27, 29]:

(1) the number of participants (registered and actual);

(2) the number of positive reviews, posted on the site;
(3) the number of events held. Their number, of course, depends on the number of grants allocated, but let us assume that the number and volume of the allocated grant also depend on the effectiveness of the implementation of activities at the University and, as a result, affect the decision-making by the Department of education.

The following is proposed as a criterion for selecting a set of those responsible for activities within a project, organized at the University [27]:

(1) The main goal is to achieve maximum interest among the project participants (it is the interest that should come first, and not the communication of new serious knowledge).
(2) At the same time, it is necessary to take into account the following circumstances: the level of competence of responsible for the events, their dynamism and teaching abilities.

Thus, in order to fulfill the above restrictions for maximizing the objective function, it is recommended that the following conditions are met [27]:

(1) There should be a certain percentage of doctors of sciences among responsible for the events (it is assumed that the qualification level of the doctor of sciences is higher).
(2) There should be a certain percentage of people with significant teaching experience among responsible for the events. It is reasonable to conclude that in order to interest the participants, the teacher must have experience of interacting with different audiences.
(3) Among responsible for the events should be people, who have experience in educational work, for example, as curators of student groups or who have experience working with schoolchildren.
(4) It is necessary to take into account the age of responsible for the event. In this case, you have to find the right balance between youth (or energy) and experience. For schoolchildren to perceive the material, it should be presented correctly, but the dynamics of the actions of the lecturer (master class, training, excursion, workshop) should not allow participants to lose attention even for a moment.

The following Tables 6.3, 6.4 and 6.5 show data for each of these items for the last three years.

The analysis of the data, presented in Table 6.3, shows that the main part of responsible for the events are teachers with scientific degrees. This is due to the fact that the level of qualification of teachers when conducting such an important project for schoolchildren with a non-standard set of events is important.

The analysis of the data, presented in Table 6.4, shows that the main support for the "University Saturdays" project is also experienced teachers. This fact is also explained by the fact that high-quality implementation of events within the framework of the project requires a high level of pedagogical skills [27, 30].

Table 6.3 Distribution of teachers with scientific degrees among responsible for events

The year of the University Saturdays conducting	Number (%) of doctors of science among teachers	Number (%) of candidates of science among teachers	Number (%) of teachers without degree
2017	5 (33)	8 (53)	2 (14)
2018	13 (42)	13 (42)	5 (16)
2019	13 (45)	11 (38)	5 (17)

Table 6.4 Distribution of teachers by teaching experience among responsible for events

The year of the University Saturdays conducting	Teaching experience more than 25 years	Teaching experience from 10 to 25 years	Teaching experience less 10 years
2017	13	2	0
2018	22	3	6
2019	17	5	7

Table 6.5 Distribution of teachers by age among responsible for events

The year of the University Saturdays conducting	Up to 40 years old	40–60 years	More than 60 years
2017	3	9	3
2018	3	9	19
2019	3	12	14

The analysis of the data, presented in Table 6.5, shows that the majority of teachers, participating in the project, are middle-aged and elderly people. This is explained both by the fact of the experience of these teachers, and by the objective circumstance, associated with the small number of young teachers at the university [27].

One of the important factors influencing the increase in the interest of participants in the project "University Saturdays" is the variety of topics of events. The main directions in which events are held at the MSTU CA are the following:

(1) Humanities. Within the framework of this direction, popular scientific lectures are held on the history of aviation and the university, about leading domestic and foreign aircraft designers, as well as the state of the world aviation system.
(2) Radar. Within the framework of this direction, participants receive information about the device of radar facilities and modern technical means, for example, optoelectronic, on the basis of which the aircraft is monitored.
(3) Avionics. Within the framework of this direction, the participant's orientation is being formed for training in the specialties "Technical maintenance of transport radio equipment", as well as "Navigation and Air traffic management".
(4) Maintenance of aircraft and aircraft engines.

(5) Ecology. Within the framework of this direction, environmental issues, related to the use of aviation, are considered.
(6) Psychological sciences. Within the framework of this direction, interesting workshops and trainings are held, and the purpose of which is to demonstrate to participants various exercises to test their ability to master the profession.
(7) Physical and mathematical sciences.

A wide range of activities, carried out within the framework of the University Saturdays project, aims to fulfill the following tasks:

(1) coverage of secondary school pupils with different interests and, as a result, improvement of such a formal indicator of the "University Saturdays" project as the number of registered and actual participants;
(2) implementation of the maximum possible career guidance for all potential applicants.

Another significant factor, affecting the interest of schoolchildren in the entire cycle of project activities, is several events that differ from others in their non-standard nature. Such events, held at the MSTU CA, include the following:

(1) excursion to the Training center of MSTU CA;
(2) excursion to the Training aviation and technical center of MSTU CA;
(3) psychological master classes.

Such events must necessarily be present in the project, since, firstly, they clearly indicate the specifics of the university, and, secondly, they arouse interest by giving the opportunity to visit "live" aviation equipment and participate in the simulation of an aircraft flight.

The interest in a particular event (its "brand") is determined by its rating. The rating can be summed up from such indicators as the number of positive reviews, the speed of electronic registration on the site, as well as such an informal indicator as the interest of parents and teachers, expressed in making inquiries by phone or e-mail.

It seems natural to evaluate the entire project "University Saturdays" in one particular university on the basis of the total rating.

We introduce a value that characterizes the rating of the event [30]:

$f_i(x, x_1, y, z), \quad i = 1, \ldots, N$, where N—the number of events, x—the number of positive reviews on event, X_1—the number of negative reviews on event, y—indicator of the speed of electronic registration on the site, and z—the number of inquiries about the event, a priori conducted by phone and e-mail.

The specifics of the project "University Saturdays", as already mentioned above, is that it should not contain events with topics that do not arouse interest among schoolchildren and, moreover, held at a low level of quality, and there must also be "locomotive events" that contribute to the increased interest of schoolchildren in the university.

Thus, it is possible to propose the following criteria for evaluating the quality of the "University Saturdays" project for a particular university:

$$\sum_{i=1}^{N} f_i(x, x_1, y, z) \to \max,$$
$$\nabla i \ f_i(x, x_1, y, z) \geq d,$$
$$\max \ f_i(x, x_1, y, z) \geq e.$$

In this criterion, the values d and e are the restrictions on the rating of each event of project and the best events.

When calculating the value that characterizes the rating of an event, the weight of individual indicators should also be taken into account. For example, the presence of one negative review is more significant for evaluating an event than several positive ones.

Thus, $f_i(x, x_1, y, z) = m_1x + m_2x_1 + m_3y + m_4z$, where m_i, $i = 1,.., 4$—the weight of individual indicators that affect the rating of the event.

References

1. Agieva MT, Malsagov MH, Ugolnitsky GA (2003) Modeling of the hierarchical structure of management of the education system. Rostov on don, p 234
2. Bespalko VP (1981) Elements of the theory of control of the learning process. M: Znanie, p 71
3. Clear J (1990) Sistemology: translate from English. M: Radio and communications, p 544
4. Rastrigin LA (1982) Modern principles of management of complex systems. Moscow: Sovetskoe radio, p 232
5. Logvinov II (1980) Simulation modeling of the educational process. Logvinov. M: Pedagogika, p 127
6. Marigodov VK, Slobodyanyuk AA (2000) Systematic approach to comprehensive training of engineers. Specialist 2000(8):34–36
7. Moiseev AM (1999) Program-target management of the development of education. Moscow: Pedagogical society of Russia, p 190
8. Novikov DA, Glotova NP (2004) Models and mechanisms of management of educational networks and complexes. M: Institute of management of education of the Russian academy of education, p 440
9. Burkov VN, Novikov DA (1997) Models and mechanisms of the theory of active systems in the management of quality of training specialists. M, p 157
10. Zhambyu M (1988) Hierarchical cluster-analysis and correspondence: Translate from the French. M: Finance and statistics, p 342
11. Karelina IG, Sobolev AB, Sorokin SO (2015) Monitoring the activities of educational organizations-the initiative of systemic changes in higher education. Part 1 Higher Educ Today 2015(6):37–46
12. Litvak BG (1982) Expert information: methods of obtaining and analysis. M: Radio and communications, p 104
13. Litvak BG (1996) Expert assessments and decision-making. M: Patent, p 284
14. Borzova AS (2017) Conceptual foundations of a model-oriented approach to forecasting and optimization of the personnel training system in the field of air transport operation. Econom Manage Manage Syst 24(2.1):188–194
15. Lvovich YE, Lvovich IY (2010) Decision-making in an expert-virtual environment. Voronezh: «Scientific book», p 140
16. Beresov AV (1989) Multicriteria optimization in examples. M: Nauka, p 128

17. Bespalko VP (1989) The components of pedagogical technology. M: Pedagogika, p 192
18. Talyzina NF (1986) Theoretical foundations of the development of a model of specialist. M: Znanie, p 112
19. Tatur YG (2004) Competence in the structure of the model of quality of training specialists. Higher Educ Today 2004(3):71–75
20. Tolmachev SA (1993) Representation and synthesis of the optimal logical structure of educational material in automated training systems. Cyber Syst Anal 1993(1):183–186
21. Chuchkalova IA Organizational bases of optimization of educational process in a multi-profile university / I.A. Kuchkarov // Optimization of study, pedagogy and professional training of students in higher education: interuniversity collection of scientific works. – Cheboksary: Chuvash state university, 1983, p. 10–20.
22. Bykov LA, Sareva ZF (1982) Methods of graphical structures in the optimization of management of study at the university. Ways of optimization of the study and pedagogical process at the university: interuniversity collection of scientific works. Barnaul: Altai state university, p 44–52
23. Sviridova NG, Korovin VM (1990) Designing academic disciplines, using a system-activity approach. Active methods of teaching and the quality of training specialists in university: interuniversity collection of scientific papers—L, p 39–41
24. Borzova AS, Tsipenko VG (2017) Multi-alternative optimization in the formation of educational resources, focused on models of professional activity. Model Optim Inf Technol 1(16). https://moit.vivt.ru/wp-content/uploads/2017/02/BorzovaZipenko%20_1_17_2. pdf (date of Accessed is 12 Feb 2018)
25. Chinyuchin YM, Chinyuchin IF(2004) Fundamentals of technical maintenance and repair of aviation equipment: a textbook. Part I Polyakova. Moscow: MSTU CA, p 81
26. Mogrunov IB, Nersesov TV, Bychkov VB (1998) About the approach to the coordination of state educational standards and educational and professional programs, based on matrices of logical relationships of study units. Problems of quality, its normalization and standards in education: collection of scientific articles—M, p 68–75
27. Zatuchny DA (2019) Ways to improve the quality of the project «University saturdays» for schoolchildren in Moscow on the example of the Moscow state technical university of civil aviation. In: Proceedings of the international symposium «Reliability and Quality», vol 2. Penza, p 111–114
28. Zatuchny DA (2011) Automation and control. Federal air transport agency, federal state educational institution of higher professional education «Moscow state technical university of civil aviation», Department of technical maintenance of radio-electronic systems of air transport. Moscow
29. Romancheva NI (2014) The task of choosing the optimal route for training a specialist. In: Proceedings of the international symposium reliability and quality, vol 1, p 53–54
30. Romancheva NI (2017) Exponential technologies in education. In the collection: Modern technologies in science and education. In: Proceedings of the II international scientific-technical and scientific-methodological conference: in 8 volumes. Ryazan state radio engineering university, p 20–22

MIX
Papier aus verantwortungsvollen Quellen
Paper from responsible sources
FSC® C105338

If you have any concerns about our products,
you can contact us on
ProductSafety@springernature.com

In case Publisher is established outside the EU,
the EU authorized representative is:
Springer Nature Customer Service Center GmbH
Europaplatz 3, 69115 Heidelberg, Germany

Printed by Libri Plureos GmbH
in Hamburg, Germany